Tourist Development

Douglas Pearce

Senior Lecturer in Geography
University of Canterbury, New Zealand

TOURIST DEVELOPMENT

Second Edition

Longman
Scientific &
Technical

Copublished in the United States with
John Wiley & Sons, Inc., New York

Longman Scientific & Technical,
Longman Group UK Ltd,
Longman House, Burnt Mill, Harlow,
Essex CM20 2JE, England
and Associated Companies throughout the world.

*Copublished in the United States with
John Wiley & Sons, Inc. 605 Third Avenue, New York, NY 10158*

This is a revised and enlarged edition of TOURIST DEVELOPMENT first
published in the TOPICS IN APPLIED GEOGRAPHY series in 1981.

This edition first published 1989

Reprinted 1991 (twice), 1992, 1994

British Library Cataloguing in Publication Data
Pearce, Douglas G., *1949–*
    Tourist development.—2nd ed.
    1. Tourist industries. Planning.
    Geographical aspects
    I. Title
    338.4′791

ISBN 0-582-01435-2

Library of Congress Cataloging-in-Publication Data
Pearce, Douglas G., 1949–
    Tourist development.

    Bibliography: p.
    Includes index.
    1. Tourist trade.   I. Title.
    G155.AIP36   1989   380.1′4591   88-37748
    ISBN 0-470-21339-6 (U.S.)

Set in 10/11pt Ehrhardt Roman

Produced by Longman Singapore Publishers (Pte) Ltd
Printed in Singapore

For Chantal and Rémi

Contents

List of Figures

List of Tables

List of Plates

Preface to the Second Edition

Preparing the second edition of this book has been a challenging exercise, involving several different tasks and considerations. A first concern has been with bringing material up to date so as to take account not only of changes in the nature and expansion of tourism in the last decade but also of the vast explosion of literature on the subject and changing attitudes and findings reflected therein. This has not necessarily meant the rejection or non-inclusion of earlier work for much can be learnt about recent research by setting it alongside the approaches, findings and attitudes of the 1960s and 1970s. What was held to be true then has been both confirmed and questioned by more recent studies as undoubtedly research from the 1980s will be re-evaluated in coming decades. Throughout the literature on tourist development there is evidence now of a growing maturity reflected in more balanced approaches to the topic, methodological refinements and some move away from ideographic studies to address questions of a more general nature.

The opportunity for a greatly expanded discussion of tourist development which publication outside the shorter, more limited format in which the first edition appeared, Topics in Applied Geography, has been welcomed and used in two different ways. Firstly, the greater length has enabled a much fuller examination of topics dealt with previously. It has allowed, for example, elaboration on the elements and agents of developments and inclusion of development processes in rural, urban and Third World areas alongside the discussion of coastal and ski-field development. Secondly, two major new themes have been introduced. In the introductory chapter tourist development is set in the broader context of the development debate. Issues raised here about what constitutes development are followed through in later chapters, particularly that dealing with the impact of tourism and the concluding chapter where tourism's contribution to development is reappraised. Tourist development too is no longer seen solely in terms of destinations, with much more consideration being given now to origins or markets and associated linkages. This change reflects not only the greater space available in the second edition but also the author's experience in the intervening period of writing *Tourism Today: a Geographic Analysis* in which the value of a more holistic approach to the study of tourism was emphasized. This approach is most evident in the new chapter dealing with demand and development but the origin–linkage–destination framework also underlies the structure of the whole book and is notable for example in those chapters dealing with agents and elements of development, impacts and planning. The new concluding chapter attempts to draw together points raised throughout the book and to anticipate the directions tourist development might take in the future. While writing the last chapter my office was shaken by a strong earthquake – whether my conclusions are earth-shattering or not I will leave to the reader's judgement.

I am grateful to many people in different parts of the world who provided feedback on the first edition or who assisted in various ways by contributing material for this new edition. In particular I am indebted to Bob Mings for his frank and constructive comments on a first draft of the new manuscript. My thanks also go to Vanessa Lawrence and the late Pam Troughton for editorial advice and assistance, to Brenda Carter, Linda Harrison and Anna Moloney for typing the manuscript, to Alister Dyer and Tony Shatford for drafting. I am especially grateful to Chantal for her continuing support and encouragement.

Doug Pearce
Christchurch
June 1988

Acknowledgements

We are grateful to the following for permission to reproduce copyright material:

the author, J B Allcock and Pergamon Press Inc for table 6.6 (Allcock 1986); The American Geographical Society for fig 3.4 (Barker 1982); the author, R Baretje and Pergamon Press Inc for table 6.5 (Baretje 1982); the British Tourist Authority for table 7.1 (BTA 1976); Butterworth Scientific Ltd for tables 6.2 & 6.7 (Lee 1987); the Canadian Association of Geographers for fig 1.4 (Butler 1980); Centre on Transnational Corporations for table 2.3 (Centre on Transnational Corporations 1982); Commission of the European Communities for table 4.4 & fig 4.5 (European Omnibus Survey 1986); the author, E E Day for fig 5.1 (Day et al 1977); the author, Jozef W M van Doorn for fig 1.2 (van Doorn 1979); the author, Professor E Gormsen for fig 1.5 (Gormsen 1981); the author, Dr N Graburn and Pergamon Press Inc for table 4.1 (Graburn 1983a); the author, Professor Clare A Gunn and Crane, Russak & Co for fig 5.4 (Gunn 1979); the author, Wolfgang Haider for table 3.2 (Haider 1985); International Development Research Centre for table 6.3 (Attanayake et al 1983); the editor for table 2.2 (ITQ 1984); the author, Dr Seppo Iso-Ahola and Pergamon Press Inc for fig 4.1 (Iso-Ahola 1982); the editor for table 2.4 (Wood 1979); the editor for tables 4.8 & 4.9 (Pearce & Johnston 1986); the author, Professor J Lundgren for fig 2.2 (Lundgren 1972); the author, Professor J Lundgren and the editor for fig 3.7 (Hills & Lundgren 1977); the author, S Milne for fig 7.2 (Milne 1987); the editor for tables 4.5 & 4.6 (Pearce 1987d); Northland United Council for fig 7.6 (Northland United Council 1986); the editor for table 5.2 (Britton 1980b); Pergamon Press for fig 4.2 (Pearce & Grimmeau 1984); Pergamon Press Inc for fig 1.3 (Miossec 1976); the author, Dr E E Rodenburg and Pergamon Press Inc for table 3.3 (Rodenburg 1980); the author, Gerda Priestley for fig 3.1 (Priestley 1986); Royal Scottish Geographical Society for fig 6.4 (Getz 1986); the editor for table 3.4 (Wong 1986); Tourism Authority of Thailand for figs 7.3 & 7.4 (TDC-SGV 1976); the editor for fig 3.5 (Lundgren 1974); the editor for fig 5.2 (Harker 1973); United Nations Environment Programme for fig 5.3 (Pearce & Kirk 1983); the author, Dr G Wall for fig 6.5 (Wall & Wright 1977); World Tourism Organisation for tables 4.2 (WTO 1983a) & 6.8 (WTO 1983c).

1

Tourism, development and tourist development

Tourist development can be defined in different ways and viewed from several perspectives but essentially it is a hybrid term embodying two basic concepts – tourism and development. This book thus begins with a review of these two concepts and the interrelationships between them to set the scene for a fuller discussion of tourist development. Such a review is important at the outset, for relatively few links have been made between the body of literature on tourism and that on development. An outline of the subsequent structure of the volume is then given.

TOURISM

Tourism has been defined in various ways but may be thought of as the relationships and phenomena arising out of the journeys and temporary stays of people travelling primarily for leisure or recreational purposes. While writers differ on the degree to which other forms of travel (e.g. for business, for health or educational purposes) should be included under tourism there is a growing recognition that tourism constitutes one end of a broad leisure spectrum. In a geographical sense, a basic distinction between tourism and other forms of leisure, such as that practised in the home (e.g. watching television) or within the urban area (e.g. going to the cinema or a walk in the park), is the travel component. Some writers employ a minimum trip distance criterion but generally tourism is taken to include at least a one-night stay away from the place of permanent residence or origin. The spatial interaction which arises out of the tourist's movement from origin to destination is an inherent and defining feature of tourism and the subject lends itself readily to geographical analysis (Pearce, 1987a).

The travel and stay attributes are also characterized by the demand for and provision of a wide range of goods and services. In terms of the tourist's destination, these can be grouped into five broad sectors: attractions, transport, accommodation, supporting facilities and infrastructure. The attractions help encourage the tourists to visit the area, transport services enable them to do so, the accommodation and supporting facilities (e.g. shops, banks, restaurants) cater for their well-being while there, and the infrastructure assures the essential functioning of all of these. Many of these services may be combined and provided by tour operators located at the destination, origin or with links to both, who offer the traveller a package involving transport, accommodation and perhaps sightseeing or some other recreational activity. Sales of these packages and/or individual travel items are often made through retail travel agencies located in the origins or markets.

There is then no single tourist product, no one commodity or service by which the output of tourism might be measured. Rather, tourists when they travel acquire an experience made up of many different parts, some tangible (transport from A to B, accommodation, souvenirs purchased), some more intangible (the pleasure of an island sunset, the thrill of white-water rafting, the appreciation of works of art, the satisfaction with high quality service in a French restaurant). Much of this experience is acquired at or *en route* to and from the destination. Souvenirs and other goods may be purchased to take home but many of the goods and services demanded by tourists are consumed *in situ*, in the places where they are offered or produced. The degree to which this occurs distinguishes tourism from many other economic activities, such as manufacturing and agriculture. Consumption at the point of production also influences the pattern of impacts which tourism may have.

Tourism is thus a multi-faceted activity and a geographically complex one as different services are sought and supplied at different stages from the origin to the destination. Moreover, in any country or region there is likely to be a number of origins and destinations, with most places having both generating (origin) and receiving (destination) functions (Pearce, 1987a).

Conceptually, many of these elements are brought together in Thurot's multiple origin–destination model. In each of the three national systems depicted in Fig. 1.1 Thurot distinguishes between supply and demand and between domestic (or internal) and international tourism. Part of the demand for tourism generated in country B, probably the larger part, will be fulfilled by that country's tourist facilities with the remainder being distributed to countries A and C. At the same time,

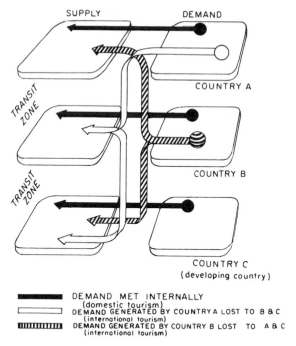

SUPPLY   DEMAND

COUNTRY A

TRANSIT ZONE

COUNTRY B

TRANSIT ZONE

COUNTRY C
(developing country)

DEMAND MET INTERNALLY
(domestic tourism)
DEMAND GENERATED BY COUNTRY A LOST TO B & C
(international tourism)
DEMAND GENERATED BY COUNTRY B LOST TO A & C
(international tourism)

Fig. 1.1   Schematic representation of the supply of and demand for domestic
and international tourism in a series of countries.
(*Source:* after Thurot, 1980).

part of the demand from country A will be channelled to country B (and
to country C), which thereby becomes an international destination as
well as a source of international travellers. In contrast, no international
demand is shown to emanate from country C, although it may generate
domestic tourists and receive tourists from countries A and C. Country
C is said to represent certain Third World countries where standards
of living may generally be insufficient to generate international tourism
(although often a small élite may indulge in a large amount of such travel)
and Soviet bloc countries where severe restrictions on international
travel may exist.

In the real world, however, the boundaries of tourism are not always
neatly confined. On the supply side, for example, facilities and services
used by tourists may be purpose-built or designed for them or they may
be shared with other users. Facilities built expressly for tourism range
from attractions such as Disneyland to resort hotels to ski-field access
roads. Others have been transformed from their original function to

some tourist use, for example farm cottages have become second homes and old canals and waterways have been restored for recreational boating. In other instances tourism may supplement or complement the original activity – wine-makers have opened their cellars to tourists and Gothic cathedrals today attract the curious as much as the faithful. Or tourists may share their accommodation and transport with other travellers and take advantage of services and infrastructure provided essentially for the resident population.

Classic expressions of tourism include coastal, alpine or thermal resorts such as Benidorm, Chamonix and Badgastein. Less distinct but no less important in terms of tourism are cities such as Tokyo, London and Paris which each year attract millions of visitors and in turn generate millions of travellers. Rural areas, where tourism is often more dispersed and its organization less formal, may also be significant tourist destinations.

Tourism has developed in many different contexts. Modern mass tourism has its origins in the affluence of the industrialized nations of Western Europe, North America and, more recently, Japan. Tourism has also expanded significantly in Eastern Europe and is becoming an important sector in many developing countries in Asia, Africa, Latin America, the South Pacific and the Caribbean. Thus tourism has developed in liberal Western societies, under centrally planned socialist regimes, as a relatively small part of large industrial economies or again as the leading sector of small developing countries. Likewise, tourism has developed in a wide variety of physical environments – on low islands of the Pacific, in the heart of alpine Europe, in the countryside of the English Lakes District and along the Mediterranean coastline.

Today there can be little doubt that tourism has assumed considerable economic and social significance throughout the world. In 1986 the World Tourism Organization (WTO) estimated the number of international tourist arrivals at about 340 million and receipts from international tourism at 115 billion dollars. Domestic tourists are generally several times more numerous than international tourists in most developed Western States.

Tourism research

Tourism has been studied by an increasing number of researchers from a variety of disciplines over the last two decades. However, no widely accepted interdisciplinary field of tourism studies has yet been defined and a cohesive body of knowledge which might be thought of as a specialized sub-discipline has been slow to emerge in geography, economics, management, sociology, anthropology and other disciplines.

Pearce (1979a) identified six broad topic areas as constituting the major components of the geography of tourism: spatial patterns of supply, spatial patterns of demand, the geography of resorts, the analysis of tourist movements and flows, the impact of tourism and models of tourist space. More recent national reviews show coverage of topics varies considerably from country to country and that the literature on tourism is still rather fragmented (Barbier and Pearce, 1984; Benthien, 1984; Duffield, 1984; Lichtenberger, 1984; Lundgren, 1984; Mitchell, 1984; Pearce and Mings, 1984; Takeuchi, 1984). General studies on the geography of tourism (Robinson, 1976; Hudman, 1980) have recently been complemented by major works on more specific areas, with studies on tourism's impacts (Mathieson and Wall, 1982; Murphy, 1985); methodological issues (Smith, 1983) and the spatial structure of tourism (Pearce, 1987a).

In economics, management and marketing, seminal works on tourism appeared comparatively early (Gray, 1970; Baretje and Defert, 1972; Burkart and Medlik, 1974; Wahab, Crampon and Rothfield, 1976), yet in the following decade the literature in these fields appears no more extensive nor cohesive. Booms and Bitner (1980), for instance, argue that tourism enterprises are strictly service enterprises and observe (p. 338) that 'with few exceptions, the management literature, management research and management training all focus on manufacturing activities and the problems of goods production'. Contributions by economists to tourism research are seen by Gray (1982) to be in the areas of measurement, cost–benefit analysis, resource allocation and the use of public goods in the development of tourism. The balance of payments effect of tourism is the main concern of the special issue of the *Annals of Tourism Research* (Vol. 9, No. 1) devoted to the economics of international tourism.

In a very useful review of the sociology of tourism, Cohen (1984) identifies eight main sociological perspectives on tourism: tourism as commercialized hospitality, as democratized travel, as a modern leisure activity, as a modern variety of the traditional pilgrimage, as an expression of basic cultural themes, as an acculturative process, as a type of ethnic relations and as a form of neocolonialism. Cohen then suggests sociological research on tourism 'falls naturally' into four principal issue areas: the tourist, relations between tourists and locals, the structure and functioning of the tourist system, and the consequences of tourism. Anthropologists, according to Graburn (1983), have also focused on the study of the impacts of tourism on host populations as well as on the study of tourists themselves. In his review, Graburn explores the notion of tourism as ritual and play.

The impact of tourism is the most common theme to emerge from these disparate studies, with varying emphasis being given to economic, social, cultural and environmental issues depending on the background of the writers concerned. Each discipline also has its own perspectives on the way in which tourism is organized, with geographers highlighting its spatial structures, sociologists and anthropologists underlining social relationships and economists emphasizing economic attributes.

DEVELOPMENT

Recent reviews of the 'meaning of development' have highlighted the many ways in which development has been used and the numerous interpretations which continue to be given to it (Goulet, 1968; Seers, 1969, 1977; D. Smith, 1977; Mabogunje, 1980; Unwin, 1983; Thirlwall, 1983; Welch, 1984). Welch (p. 2) considers development 'has become a term bereft of precise meaning' and quotes Friedmann (1980) who suggests it 'is one of the more slippery terms in our tongue'. That there is no single, unequivocal definition of development is due in part to different uses of the term by different disciplines and changes in those uses over time, particularly in the last three decades.

Much of this ambiguity results from the use of the term 'development' to refer both to a process and to a state. Goulet (1968), one of the earlier writers in the field, notes (p. 388):

> Development has usually been treated as a process, a particular kind of social change. Nevertheless, development is also a state or condition. Whenever a society is called developed or underdeveloped we refer to its present condition. Similarly, when development is declared to be a major goal of Third World nations, the allusion is to a terminal condition, not to a process. Thus the single term 'development' refers both to the destination of a journey and to the journey itself.

Development, according to Friedmann (1980, p. 4):

> suggests an *evolutionary* process, it has positive connotations . . . And of course, development is always *of* something particular, a human being, a society, a nation, an economy, a skill . . . It is often associated with words such as *under* or *over* or *balanced*: too little, too much, or just right . . . which suggests that development has a structure, and that the speaker has some idea about how this structure *ought* to be developed. We also tend to think of development as a process of change or as a complex of such processes which is in some degree lawful or at least sufficiently regular so that we can make intelligent statements about it.

Rostow (1960), in one of the most widely cited (though not universally accepted) examples of development as process, claimed to identify five successive stages of economic growth: traditional, transitional, take-off, maturity and high mass consumption. Moreover, he suggested societies or nations could be classified according to these stages with the implication that there was a natural path to economic growth which nations follow.

Much of the literature on development as a state has focused on how to measure that state. D. Smith (1977, p. 203) notes that 'development is frequently assumed to be an economic condition' and that 'the most common measure of development is an economic indicator – GNP per capita'. Debate in this area has revolved around whether such measures are adequate indicators of development and if not, as is being increasingly recognized, what other social indicators might complement or replace them. This has given rise to quite an extensive literature on socio-economic indicators involving such measures as infant mortality rates and levels of protein intake.

There have also been calls for less technical definitions of development. D. Smith (1977) argues for development as 'welfare improvement' and suggests (p. 207) 'development means a better state of affairs, with respect to who gets what where.' Earlier, Goulet (1968) had identified as three major goals of development – sustenance of life, self-esteem and freedom – each broadly defined.

This broadening of the concept of development, both as process and state, away from narrow considerations of economic growth to encompass wider economic then social concerns has contributed significantly to the burgeoning range of definitions of the term in the last two decades. This is not to say that earlier definitions have been completely abandoned. Development is still seen solely in terms of economic growth in some quarters.

As an illustration of the shift in the concept of development, Mabogunje (1980) cites two of the seminal papers by Seers (1969, 1977). In the first, Seers writes:

> The questions to ask about a country's development are three: What has been happening to poverty? What has been happening to unemployment? What has been happening to inequality? If all three of these have declined from high levels, then beyond doubt there has been a period of development for the country concerned.

Eight years later Seers (p. 5) underlines the importance of an additional element, self-reliance:

> On this approach, 'development plans' would henceforward not put the main emphasis on overall growth rates, or even on new patterns of distribution. The crucial targets would be for (i) ownership as well as output in the leading economic sectors; (ii) consumption patterns that economised on foreign exchange . . . (iii) institutional capacity for research and negotiation; (iv) cultural goals [e.g. reducing cultural dependence on one or more of the big powers].

Mabogunje then goes on (pp. 36–46) to identify four main ways in which the term development has been used, before introducing a fifth definition of his own. These are:

7

1. Development as economic growth

Mabogunje argues that in the early post Second World War period, development was interpreted narrowly in terms of economic growth with priority given to 'increased commodity output rather than to the human beings involved in the production.' A common expression of this in the underdeveloped countries was concentration on export production and the emergence of a dual economy.

2. Development as modernization

Later a social dimension was incorporated. 'Development, still in the sense of economic growth, came to be seen as part of a much wider process of social change described as modernization. . . . The emphasis on development as modernisation is . . . on how to inculcate wealth-oriented behaviour and values in individuals.' Education was seen to be a critical aspect of societal change but modernization also had a consumption dimensions: 'To be modern meant to endeavour to consume goods and services of the type usually manufactured in advanced industrial countries.'

3. Development as distributive justice

By the late 1960s attention was being increasingly turned to who was getting, or not getting, the benefits of social and economic change:

> Development came to be seen not simply as raising per caput income but more important, of reducing the poverty level among the masses . . . interest in development as social justice . . . brought to the forefront three major issues: the nature of goods and services provided by governments for their populations; the question of accessibility of these public goods to different social classes; and the problem of how the burden of development (defined as externalities) can be shared among these classes.

The latter factor is an important extension of the concept, incorporating as it does not only who benefits but also who pays for development in terms of such externalities as air and water pollution resulting from industrialization. Regional development planning emerged as an important strategy for distributional justice.

4. Development as socio-economic transformation

This interpretation is attributed by Mabogunje to 'scholars of a Marxist philosophical persuasion [who] argue that the questions of distribution and social justice cannot be resolved independently of the prevailing mechanisms governing production and distribution.' This interpretation is essentially a critique of the capitalist 'mode of production' (those elements, activities and social relationships which are necessary to produce and reproduce real (material) life). 'Basic shifts in any of the aspects of the mode of production can trigger off wide-ranging changes which may culminate not only in the transformation of the mode but also in changes in the relative importance of social classes. It is such a socio-economic transformation that really constitutes development'. This interpretation stresses the interrelationships between development and underdevelopment with developed metropolitan centres becoming enriched at the expense of the 'underdeveloped' peripheral regions. As such, it is closely linked to dependency theory.

5. Development as spatial reorganization

Mabongunje himself stresses (pp. 65–8) the spatial dimension of development: 'spatial reorganization is seen as synonymous with development in the sense that spatial forms represent physical realizations of patterns of social relations'. The need for a pattern of social relations which can inculcate new processes of production thus requires the reconstruction of spatial structures both in the rural and urban areas of a country. Mabongunje raises the notion 'that certain types of spatial arrangement can be expected to make a relatively better contribution to the attainment of specified goals than others', and underlines the importance of magnitude and the time factor in development.

The development literature discussed here has largely focused on the underdeveloped countries. Development, nevertheless, does still occur in the developed world. There many of the issues raised above have been discussed within the context of the more specific literature on regional development. Yuill, Allen and Hull (1980), for example, suggest there are three distinct phases of regional policy in Europe. In the first, immediately after the Second World War, the emphasis was on national economic growth with (p. 217) 'an increase in the national cake being viewed as far more important than questions of how the cake might be spatially distributed'. A second period, from the late 1950s to the early 1970s, is characterized as the heyday of regional development, for with growing economic prosperity came a concern for greater social justice and an evening out of regional inequalities which an increasing number of studies using socio-economic indicators had identified. In the third, most recent period, that following the general downturn in many European economies since the energy crisis of 1973–74, attention has turned once again to the size of the national cake rather than the manner in which it is distributed. As Yuill *et al.* (p. 221) put it, 'in short, for regional incentive policy the post-1973 period has been very much one of consolidation in a hostile environment'.

In the case of France, these different phases can be seen in the changing emphases of the various Five Year Plans of the post-war period. The first three plans (1947–53, 1954–57, 1958–61) were essentially aimed at reconstruction and modernization of key industries. The Fourth (1962–65), Fifth (1966–70) and Sixth (1970–75) Plans became 'Plans for Economic and Social Development'. A social and regional dimension was incorporated for the first time in 1962. However, by the Sixth Plan, economic growth was once again dominant but 'social priorities were not to be sacrificed to the needs of economic progress' (House, 1978, p. 24). The Seventh Plan (1976–80) was characterized by a move to smaller scale projects under the Priority Action Programmes while the Eighth (1981–85) reflected the Socialists' moves towards decentralization.

Development then takes on many meanings, being used to refer both to a process and to a state and to have a range of defining characteristics from narrowly delineated economic ones through broader social values to more general attributes such as self-reliance. Generally the change has been away from narrow economic usage to embrace other attributes. The links between process and state suggested by Goulet (1968) have, however, not always been explicitly made, that is that the state of development derives from the processes which have caused it.

TOURISM AND DEVELOPMENT

The development literature generally ignores tourism and few writers on that subject set their studies in the broader context of development although they may address some of the specific questions raised in the previous section. None of the reviews of the development literature discussed above, for instance, alludes to tourism despite its growing economic and social significance and use in development strategies in many developing countries over the last three decades. In this respect tourism has been treated by development writers in the same way they have ignored other service sectors for much of the development debate has centred on the transition from an agricultural to an industrial society and neglected tertiary activities.

Exceptions to this pattern occur nevertheless, as in several discussions of underdevelopment in the Mediterranean. Schneider, Schneider and Hanson (1972), for example, take tourism as one aspect of modernization, citing the emulation by locals of tourist behaviour as an illustration of the 'incorporation of distorted metropolitan lifestyles by dependent regions and issues of economic dependence'. For Seers (1979), the centrifugal flow of tourists from Western Europe to countries of the southern Mediterranean is one of the manifestations of the core–periphery relationship between the two regions.

Direct links between tourism and theories of economic development were made in several early tourism papers which have been quoted rarely or have had limited circulation (e.g. Krapf, 1961; Kassé, 1973; van Doorn, 1979) as well as in a couple of tourism books which have been widely cited (Bryden, 1973; de Kadt, 1979). A small number of more recent writers have also addressed these issues squarely (Britton, 1982; Erisman, 1983).

In his pioneer paper, Krapf (1961) raises a number of explicit questions. Which are the developing countries? What is the nature of economic growth? What types of aid can be given to these countries? What role can tourism play? Krapf draws heavily on Rostow's model and notes in passing that mass tourism is one consequence of an age of mass consumption. He then goes on to ask whether tourism is an appropriate form of aid; whether the development of a luxury sector is justified when many people do not have the bare essentials of life. Krapf concludes that tourism has a 'special function' in developing countries, a function which he defines in terms of a series of 'economic imperatives', viz:

- exploitation of the countries' own natural resources,
- international competitiveness due to favourable terms of trade,
- an ability to provide internally many of the goods and services required,
- improved balance of payments,
- social utility of investments in tourism: employment generation and multiplier effect,
- balanced growth.

Krapf's emphasis is clearly on tourism's contribution to economic growth and the notion that tourism had a special function in this regard appears to be one which was widely held in the 1960s and one which has persisted.

By the early 1970s Kassé (1973) is writing of the emergence of a 'theory of the development of the tourist industry in under-developed countries'. This he summarizes in terms of tourism's perceived ability to generate, from limited investment in plant and infrastructure, large sums of capital which may be transferred to other sectors of the economy. The theory also underlines the multiplier impact of tourism, the creation of employment, public revenue and foreign exchange earnings. Kassé, however, does not take these points for granted but raises two basic questions. What are the real costs of developing tourism? What are the effects, direct and indirect, which tourism has on the rest of the economy? While acknowledging the number of variables which much be taken into account and deficiencies in the data needed to answer these questions, Kassé, drawing on the African experience, suggests the costs of tourism may be greater and the benefits smaller than popularly supposed. Kassé also broadens the economic argument and discusses some of the social implications of tourist development, citing in particular the work of Ben Salem (1970) in Tunisia.

In a larger, better known study from the Caribbean entitled *Tourism and Development*, Bryden (1973) raises similar issues and attempts to document them more fully. Bryden (p. 218) concludes that 'To state that this study has provided a definitive economic case against the further development of tourism in the Caribbean would be going too far', but continues that it does 'raise some very serious doubts about the viability of tourist development *in its present form*'. This latter point is an important one as Bryden was one of the first to recognize explicitly that tourism development takes different forms and its impact is conditioned by the context in which that development occurs. In particular, he notes that his general conclusions about the viability of tourism in the Caribbean result from the high degree of foreign ownership and consequent repatriation of profits, the employment of non-nationals with similar results and the real costs to the nation of government involvement in the provision of infrastructure and incentives.

Van Doorn (1979) takes this notion further and argues (p. 5) that 'tourism cannot be considered outside the context of the different stages of development countries have reached'. He then proposes a typology (Fig. 1.2) which combines levels of social and economic development based on prosperity and welfare criteria with levels of tourist development derived from the social impact work of Forster (1964) and Greenwood (1972). This potentially very useful typology unfortunately is not elaborated on nor illustrated by empirical examples. It also begs the question of the relationship between the stage of tourist development and the level of socio-economic development. What did tourism contribute to the country's level of social and economic prosperity? To what extent is the stage of tourist development dependent on more general economic and social conditions?

Van Doorn does, however, go on to suggest that theories of development must be taken into account in the assessment of the impact of tourism (p. 6):

> Any effect of tourism, be it raised employment, decreased awareness of cultural identity or the contribution of health facilities must be weighed *explicitly* against the underlying theory of development. Most studies have only an *implicit* notion based on an intuitive feeling of one of the two mainstreams in development theory: the *traditional* and the *modern* theory.

Van Doorn sees the traditional theory being based on 'the creation of a Western type of society as a result of a combination of economic, social and cultural changes'. Such changes are conditioned by the internal characteristics of the country in question and the solution to underdevelopment is seen to lie in economic injection or in social changes of

Stages of tourist development

STAGES OF DEVELOP-MENT		Stage I	Stage II	Stage III
	H_WL_P			
	L_WH_P			
	L L			

Key

Stage I   :   Discovery: new areas found by drifters

Stage II   :   Local response and initiative: supply stems mainly from local resources. Decisions regarding tourism, incrementally made by local authority.

Stage III   :   Institutionalization. Decisional control and tourism development passed out of hands of local community. Standardization, policy making and planning by regional or national authority.

H_WL_P :   high welfare, low prosperity

L_WH_P :   low welfare, high prosperity

L L   :   low welfare, low prosperity

Source: van Doorn 1979.

Fig. 1.2   A comparison of development stages of countries with stages of tourist development
(*Source:* after van Doorn, 1979).

roles, norms and values. Developing tourism for the reasons outlined by Krapf (1961) is consistent with this theory. For van Doorn, the 'leading lady' amongst the modern theories is dependency or core–periphery theory which explains differences in levels of economic development in terms of external factors. Here the 'central thesis is that developed countries grow autonomously while the underdeveloped countries show a growth-pattern that has been derived from the developed countries'. Examples supporting this concept are drawn by van Doorn from the literature, for example Bryden (1973), but they have generally not been expressed directly in these terms.

Several geographers have been much more explicit in setting their analysis of tourism in the context of modern theories of development. According to Hills and Lundgren (1977, p. 256) 'from the viewpoint of geographical theory, a major characteristic of international tourism is the centre–periphery syndrome . . . The periphery is . . . relegated to a subordinate function of the centrifugal process, bringing not only visible physical commodities, in the form of tourists, but simultaneously injecting powerful, and more subtle, hierarchical dimensions.' Britton's (1982, p. 332) intention is very direct, that is 'to place the study of tourism firmly within the dialogue on development'. This dialogue is equated with dependency theory (pp. 333–4). 'Dependency can be conceptualized as a process of historical conditioning which alters the functioning of economic and social sub-systems within an underdeveloped country. This conditioning causes the simultaneous disintegration of an indigenous economy and its reorientation to serve the needs of exigenous markets . . .'

One of these is the market for international tourism. After examining the organization of international tourism and case studies from the Pacific, Britton concludes (p. 355) that 'The international tourist industry, because of the commercial power held by foreign enterprises, imposes on peripheral destinations a development mode which reinforces dependency on, and vulnerability to, developed countries.'

Political scientists have examined other aspects of tourism and dependency theory, notably in the Caribbean. Francisco (1983) investigated relationships between levels of American tourist activity and political compliance amongst Caribbean and Latin American nations and concluded (p. 374): 'Economic reliance on tourism . . . may result in distorted economic development, foreign economic leakage, domestic social dissatisfaction, and resentment, but it does not result in political compliance at the international level.' Questions of tourism and cultural dependency are addressed by Erisman (1983) who recognizes (p. 350) that 'the literature contains only implicit theories based on scattered, often random comments'. The four most common implicit theories – Trickle Down, Commodization, Mass Seduction and Black Servility – are then outlined and illustrated by reference to a variety of studies from the Caribbean. Limitations with each are found with Erisman concluding that a Commodization/Mass Seduction synthesis is likely to be the most productive avenue for future research.

Other social and cultural aspects of tourism and development are drawn together by de Kadt (1979) in *Tourism, Passport to Development?* After outlining changing attitudes to development and stressing the importance of social and cultural issues he observed (p. xv): 'These social changes, together with important material effects on employment and income, are, of course, precisely the results that determine whether the process of tourism development is judged good or bad by the people affected.' De Kadt also recognizes that tourism takes various forms and is one of the few to acknowledge explicitly that the papers in his volume deal essentially with one particular form, namely resort tourism.

TOURIST DEVELOPMENT

The writers who address directly the relationships between tourism and development in the manner outlined in the preceding section are by and large exceptions. Many of the studies which touch on issues raised in the previous two sections have been phrased in terms of the 'impacts of tourism', the area identified earlier as the most extensive in the literature on tourism. Such studies commonly examine issues of tourism-generated revenue or employment, social changes induced by the expansion of tourism or the environmental impacts of tourist projects. These issues are seen in terms of the impacts of tourism but such impacts, as van Doorn (1979) has noted, are not usually set in any broader context of development, however defined. For a variety of reasons, most studies also only focus on a limited range of impacts and few pretend to be exhaustive or comprehensive. Moreover, these impacts are often divorced from the processes which have created them. Many writers speak of the impact of tourism without considering the type of tourism concerned or the way in which tourism has developed. Summing up the 1976 symposium on tourism and culture change held by the American Anthropological Association, Nash (cited by Smith, 1977a, p. 133) noted: 'In these papers generally, the causal agent (tourism or some aspect of tourism) tends not to be well delineated or explicated. We have to dig to find out what it is.'

Finding out about tourism, the way in which it develops and the effects of that development is what this book is about. It focuses on development both as a process and as a state, the focus broadening with the passage from process to state. That is to say, in terms of tourist development as a process, the emphasis is on the way in which tourism develops or evolves. Some account must be taken of more general processes of development but the focus is sectoral and the sector examined is tourism. In this respect tourist development might be narrowly defined as the provision or enhancement of facilities and services to meet the needs of tourists. Tourism, however, might also be seen as a means of development in a much broader sense, the path to achieve some end state or condition. It is in this light that the so-called impacts of tourism are re-examined. To what extent do these impacts contribute to the development of countries, regions or communities? Furthermore, what are the relationships between process and state? To what extent does tourism's contribution to the state of development in any area depend on the way tourism there has developed?

This book attempts to answer these and other questions by examining systematically different aspects of tourist development. By identifying the various factors involved and analysing the general relationships amongst them an attempt is made to provide readers with a general appreciation of the subject and a basic framework and methodology with which they may then address particular problems. Many of the basic relationships to be explored and the questions to be examined are outlined or raised in the following review of models of tourist development.

Models of tourist development

Miossec's (1976, 1977) model (Fig. 1.3), which depicts the structural evolution of tourist regions through time and space, remains the clearest and most explicit conceptualization of the process of tourist development. In Friedmann's (1980) terms, it is both evolutionary and has a well defined structure. Miossec stresses changes in the provision of facilities (resorts and transport networks) and in the behaviour and attitudes of the tourists and the local decision-makers and host populations. In the early phases (0 and 1) the region is isolated, there is little or no development, tourists have only a vague idea about the destination while the local residents tend to have a polarized view of what tourism may bring. The success of the pioneer resorts leads to further development (Phase 2). As the tourist industry expands, an increasingly complex hierarchical system of resorts and transport networks evolves while changes in local attitudes may lead to the complete acceptance of tourism, the adoption of planning controls or even the rejection of tourism (Phases 3 and 4). Meanwhile the tourists have become more aware of what the region has to offer, with some spatial specialization occurring. With further development, Miossec suggests it is tourism itself rather than the original attractions which are now drawing visitors to the area. This change of character induces some tourists to move on to other areas.

The phases of tourist development in Fig. 1.3 are not purely hypothetical as empirical studies have shown different tourist regions throughout the Mediterranean might be equated with the stages of maturity depicted in Miossec's model (Pearce, 1987a). The coast of Provence has undoubtedly reached a very mature stage, being well developed or even saturated for virtually its entire length, with a dense and well integrated transport network being established and tourism coming to shape or even dominate the urban and economic structure of the region. Market specialization occurs, distinct variations in types of accommodation are found and a hierarchy of resorts and urban centres is apparent. Languedoc–Roussillon, the Costa Brava, Costa Blanca and Costa del Sol are at an earlier stage of development. A range of resorts has been developed in each case, the communications network has been expanded to include regional motorways and, especially in the Spanish case, international airports and an overall regional structure is starting to emerge. The Mediterranean coast of Morocco is at an even earlier

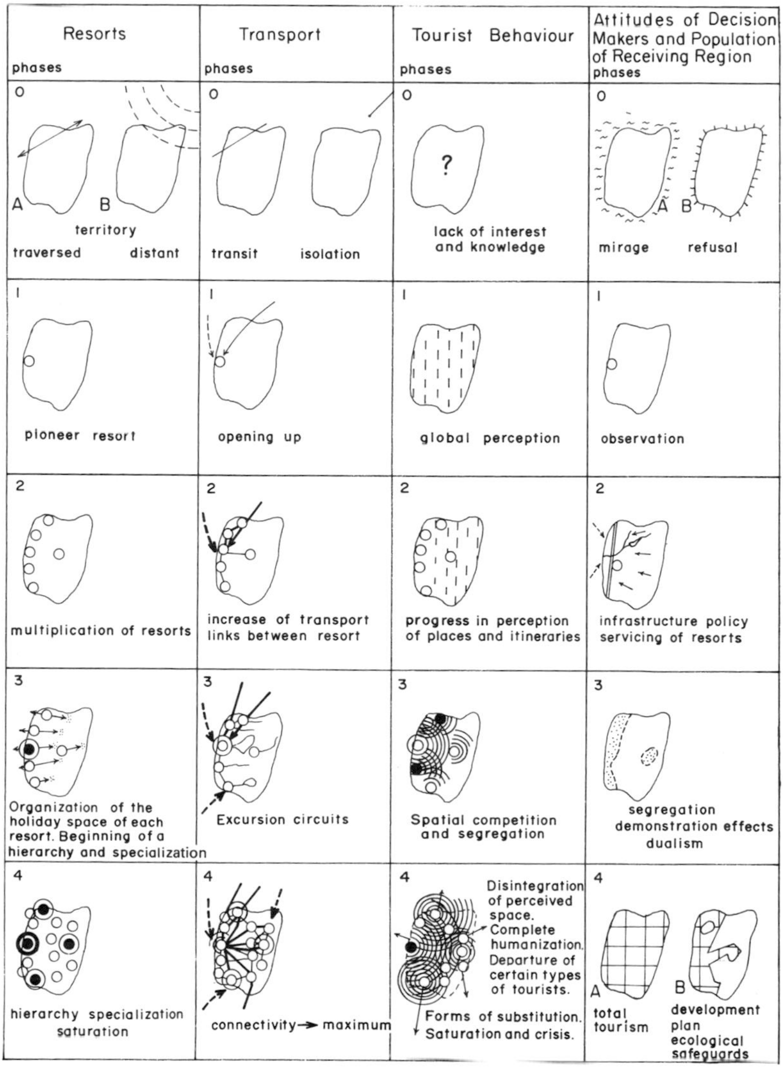

Resorts phases	Transport phases	Tourist Behaviour phases	Attitudes of Decision Makers and Population of Receiving Region phases
0 A B territory traversed distant	0 transit isolation	0 ? lack of interest and knowledge	0 A B mirage refusal
1 pioneer resort	1 opening up	1 global perception	1 observation
2 multiplication of resorts	2 increase of transport links between resort	2 progress in perception of places and itineraries	2 infrastructure policy servicing of resorts
3 Organization of the holiday space of each resort. Beginning of a hierarchy and specialization	3 Excursion circuits	3 Spatial competition and segregation	3 segregation demonstration effects dualism
4 hierarchy specialization saturation	4 connectivity→ maximum	4 Disintegration of perceived space. Complete humanization. Departure of certain types of tourists. Forms of substitution. Saturation and crisis.	4 A B total development tourism plan ecological safeguards

Fig. 1.3 Miossec's model of tourist development
(*Source:* after Miossec, 1976).

stage, perhaps Phase 3 of Miossec's model. The number of resorts there has multiplied as package tourism expanded but a coherent, hierarchical regional structure has yet to develop and there are indications that this may not occur given the difficulties experienced by some of the isolated tourist enclaves there.

As a general framework of tourist development, Miossec's model contains several useful points. Firstly, it embodies a dynamic element, the development of the region through time and space. This notion of spatial/temporal evolution is critical, both in analysing past processes and in planning the path future development is to take. Secondly, it attempts an overview of this evolution; changes in the behaviour of the tourists and the local population are related to the growth of resorts and the expansion of the transport network. However, as Miossec notes, each of the four elements need not develop apace and therein lies the source of many of the problems to which tourism may give rise. The key factor is that impact is related to development and, more importantly, particular impacts are related to specific stages of development. Other aspects of the development process are less explicit although they might be incorporated into the model. Some activities are attributed to the local population, for example the provision of supplies and the development of infrastructure, but the actual means of, and the agents for development, are not elaborated on. Who builds the resorts, how, for what reasons and with what results are fundamental questions which must be asked and answered. Likewise, the factors which influence the location of the resorts and the form of the hierarchies which emerge must also be examined. More generally, the context in which this development takes place is also neglected here as is the case of many other models. Tourism has developed in comparatively empty areas such as the Mediterranean coast of Morocco (Berriane, 1978) and developers have actively sought non-settled areas for the construction *ex nihilo* of ski resorts (Pearce, 1978a) but tourism usually develops within an existing socio-economic structure where some form of urban hierarchy and some transport networks are already found.

Other models place greater emphasis on several of the points which Miossec has not explored, notably the interrelated questions of the extent of local/non-local participation in the development process and changes in the volume and composition of the tourist traffic over time.

Figure 1.2 suggests a general decrease in local participation over time as control of and involvement in the development process passes to regional and national authorities and developers. Drawing on some of the same literature as van Doorn (1979), Stansfield's (1978) study of Atlantic City and Plog (1973) and Cohen's (1972) work on tourist typologies, Butler (1980) has developed a more complex model of the hypothetical evolution of a tourist area (Fig. 1.4).

Six stages are identified in this evolutionary sequence which is based on the product cycle concept: exploration, involvement, development, consolidation, stagnation and rejuvenation or decline. No specific

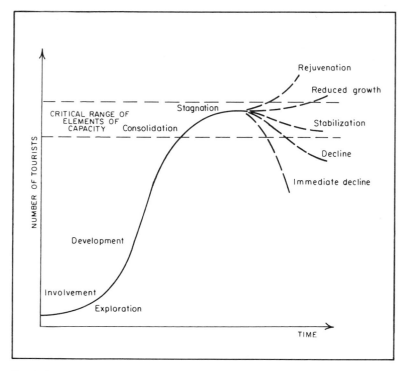

Fig. 1.4 Butler's model of the hypothetical evolution of a tourist area (*Source:* redrawn from Butler, 1980).

facilities for visitors exist in the first stage, those in the involvement stage are provided primarily by locals but then local involvement and control declines rapidly in the development phase (p. 8):

> Some locally provided facilities will have disappeared, being superseded by larger, more elaborate and more up-to-date facilities provided by external organizations, particularly for visitor accommodation. Natural and cultural attractions will be developed and marketed specifically and these original attractions will be supplemented by man-made imported facilities. . . . Regional and national involvement in the planning and provision of facilities will almost certainly be necessary and, again, may not be completely in keeping with local preferences.

Major franchises and chains in the tourist industry will be represented by the consolidation stage. Local involvement only increases again in the decline stage 'as employees are able to purchase facilities at significantly lower prices as the market declines' (p. 9).

In Gormsen's (1981) model of the spatio-temporal development of international seaside tourism, regional participation in the development process is shown to increase not decrease over time (Fig. 1.5). Gormsen's model is based on a study of the historical development of coastal tourism, essentially from a European perspective. Thus the first periphery refers to resorts on both sides of the Channel as well as those of the Baltic; the second incorporates the coasts of Southern Europe; the third includes the North African littoral and the Balearic and Canary Islands; while the fourth periphery embraces more distant destinations in West Africa, the Caribbean, the Pacific and Indian Oceans, South East Asia and South America. Approximate dates are given for the various stages in the development of each periphery. Thus unlike the models of Miossec, Butler and van Doorn, Fig. 1.5 is time and place specific, although at a fairly general scale. Gormsen also suggests comparable peripheries can be identified for the USA.

According to Gormsen's model, the initiative in the early stages comes from external developers but with time there is a growing regional participation (column A). Gormsen argues that at any given time the outermost periphery will be dependent on 'the leading strata of the metropolises' not only to generate the demand but also to develop the facilities, which are usually of a high order. Historically, in the first periphery 'the bourgeoisie not only formed part of the tourist élite but also invested in the luxurious palace hotels which were rapidly being built on cliff-tops and along promenades'. Villas for the upper classes were an important part of the initial development of the second periphery. The more recent development of the outer peripheries has also seen a dependence on external capital and international investment, much of this tourism being characterized by large-scale hotels and group tours which longer distances and travel to more alien cultures encourage.

Over time, however, changes in demand occur as participation in holidaytaking is extended to other social groups (column B) through more general processes of socio-economic transformation. This may also induce increased regional participation in the development of tourism (p. 162):

> In the peripheries closer at hand, which were opened up at an earlier
> stage, the proportion of the middle and lower classes among the seaside
> holidaymakers is . . . increasing. This is also true of the regional
> participation, as with immigration and a certain multiplier-effect the
> population is gradually becoming involved in independent activities, which
> are however limited chiefly to secondary, less profit-making tourist services.
> In some of the countries concerned, general trends in the socio-economic
> change are bringing about the formation of a middle-class which now too
> is participating in seaside tourism in its own country, in which, alongside
> financial resources and fixed holiday allocations, the adoption of spare-time
> pursuits from industrialised societies is of great importance. Examples in this

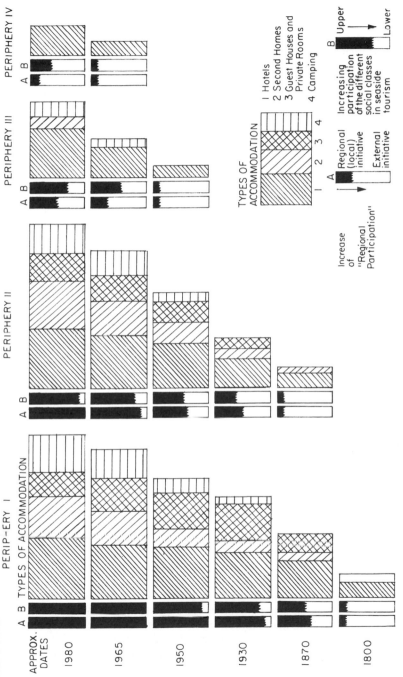

Fig. 1.5 Schematic representation of the spatio-temporal development of international seaside tourism (*Source:* redrawn from Gormsen, 1981).

category are not only from Spain and other southern European countries but also Mexico where the domestic tourism in many seaside resorts is taking over from the international.

Thurot (1973) sees the early development of international tourism in the Caribbean in terms of class succession. He outlines a process, based on the analysis of the evolution of airline routes, in which the different destinations pass through three successive phases:

Phase 1: Discovery by rich tourists and construction of an international class hotel.

Phase 2: Development of 'Upper Middle Class' hotels (and expansion of the tourist traffic).

Phase 3: Loss of original value to new destinations and arrival of 'Middle Class' and mass tourists.

Plog (1973), on the other hand, emphasizes not class but the personalities of different types of travellers. From a series of motivational studies, initially of flyers and non-flyers, Plog suggests that travellers are distributed normally along a continuum from psychocentrism to allocentrism. At the one extreme are the 'psychocentrics' who tend to be anxious, self-inhibited, non-adventuresome and concerned with the little problems in life. In contrast, the 'allocentrics' are self-confident, curious, adventurous and outgoing. Travel, according to Plog, is a way for them to express their inquisitiveness and curiosity. The travel interests and demands of the two groups differ so that different groups of travellers will visit different destinations. Moreover, Plog suggests that the market for a given destination evolves and that the destination appeals to different groups at different times. The destination will be 'discovered' by 'allocentrics', but as it becomes more well known, develops and attracts more visitors, for example the 'mid-centrics', it will lose its appeal and the 'allocentrics' will move on. As the population is said to be normally distributed, this means that an area will receive the largest number of visitors when it is attracting the 'mid-centrics', that is at a stage when it is neither too exotic nor too familiar. But from this point on, the implication is that the market will decline. According to Plog (p. 16), 'we can visualize a destination moving across the spectrum, however gradually or slowly, but far too often inexorably towards the potential of its own demise. Destination areas carry with them the potential seeds of their own destruction, as they allow themselves to become more commercialized and lose their qualities which originally attracted tourists'. Thus Butler (1980) associates the growth in the tourist traffic during the development phase of Fig. 1.4 with a broadening of the market base due to the attraction of mid-centrics while in the stagnation phase destinations will draw more on psychocentric visitors.

Reime and Hawkins (1979) argue that decline is not inevitable and might be avoided by more appropriate marketing (p. 74): 'By defining the needs of each market segment and translating these needs into elements of a touristic experience, we can prevent those forms of development that incorporate in their very design the decline of the attraction.' But marketing alone is not enough (p. 74): 'to be successful, tourism development must correspond to the inherent characteristics and needs of the region, its society and the customers sought'. Three factors must be taken into account according to Reime and Hawkins: the consumer, the producer and the society at large (p. 68):

> The history of tourism development has shown that the three elements are equally important and that long-range objectives cannot be achieved if one element is continually subordinated to the others. Examples of unstructured tourism development abound and the problems encountered frequently outweigh even the most attractive benefits. Social problems emerge as local residents – forced to serve as a cheap labour supply – revolt against the attitudes of the clientele whose demands and behaviour are often foreign to their own way of life. Ecological problems resulting from inadequate infrastructure and planning are also inherent in these unstructured, one-dimensional tourism developments.
> A successful tourism development is one in which the attraction serves as a facility for both residents and visitors. The long-lived, carefully conceived development does not force the whims and aspirations of a multitude of strangers on a region – it uses the indigenous qualities of the region, whether social or natural, to satisfy the expressed needs of a selective clientele.

Reime and Hawkins then propose a set of criteria for selecting desirable development alternatives:

- is it economically viable?
- is it socially compatible?
- is it physically attractive?
- is it politically supportable?
- is it complementary?
- is it marketable?

With the addition of a further criterion – is it environmentally sustainable? – this might be considered a very useful general set of questions to ask about tourist development.

SUMMARY

The different models and concepts reviewed here are by no means all-embracing and the field of tourist development is yet to be fully supported by a strong theoretical base. Many of the hypotheses incorporated in these models conflict, competing explanations of processes are given and there has been little empirical testing of the ideas and theories advanced. Nevertheless, the review of the models of tourist development, together with the earlier discussions of tourism and development, have raised a series of important issues, many of which are examined further in subsequent chapters.

Building on the basic framework provided by Miossec's model (Fig. 1.3), Ch. 2 examines in more detail the various elements involved in tourist development, outlines the roles and functions of the various agents of development, and discusses the importance of the context of development. These factors are then drawn together in a discussion of various typologies of tourist development (Ch. 3). Aspects of demand and supply (Fig. 1.1) will be examined in the following two chapters. Chapter 4 will review the motivations for tourist travel and discuss the factors affecting demand and how they relate to development. This is complemented by Ch. 5 which analyses the factors influencing the distribution of tourist facilities and site selection before discussing methods of assessing tourism potential. General frameworks for assessing the impact of tourism are presented in Ch. 6 and followed by an examination of the more specific economic, social and environmental impacts which tourism may have and the means of evaluating these. These impacts are interpreted as the results of different processes of development and in terms of their contribution to the state of development in the areas concerned. In the light of these impacts and the processes outlined earlier, Ch. 7 considers planning for tourism at the national, regional and local levels. Throughout these chapters general points and principles will be illustrated by a geographically diverse range of examples, as such an approach allows a fuller evaluation of the generality of patterns, processes and problems and the incorporation of a broader spectrum of concepts and techniques. The book concludes with a critical review of processes of tourist development and tourism's contribution to development and considers factors likely to influence the path tourist development is likely to take in the future.

Elements, agents and the context of development

The multi-faceted nature of tourism was highlighted in Ch. 1. Tourists seek a wide range of facilities and services which are often provided by a multitude of different suppliers at different stages of the trip or vacation. This chapter outlines the spectrum of supply, considers the roles and functions of the various agents of development and stresses the importance of taking into account the context in which tourist development occurs. Although by no means exhaustive, this structured examination aims to provide a general basis for analysing processes of tourist development and an introduction to the review of different typologies which follows in Ch. 3.

ELEMENTS OF SUPPLY

If tourism is seen in terms of a market (origin)–linkage–destination system (Pearce, 1987a) then many of the services and facilities sought by tourists are found at the destination. For this reason, much of the tourist development literature is destination-oriented with attention being focused on such sectors as attractions, accommodation, supporting facilities and infrastructure. The transport sector provides the essential link between markets and destinations with travel also occurring within each of these zones. Many of the wholesale and retail travel functions, however, are to be found in the market where tour operators and retail travel agencies put together, promote and sell individual or packaged items of the tourism experience.

Attractions

Many different attractions may induce tourists to visit particular areas or spend their holidays in specific regions. These have been classified in a variety of ways (Suzuki, 1967; Peters, 1969; Defert, 1972). A first distinction is usually made between natural features such as land-forms, flora and fauna and man-made objects, historic or modern, in the form of cathedrals, casinos, monuments, historic buildings or amusement parks. A third general category embraces man and his culture as expressed through language, music, folklore, dances, cuisine and so forth. Lew (1987), in a recent review of studies of attractions, suggests three basic perspectives have been adopted: an ideographic listing of attractions, an organizational perspective which takes account of factors such as capacity, spatial and temporal scale, and a cognitive perspective incorporating tourist perceptions and experiences of attractions. More specific attractions and means of evaluating them are discussed in Ch. 5.

Accommodation

Many different forms of accommodation are available to the modern tourist. These might be broadly classified into the commercial sector (hotels, motels, guest-houses, holiday camps, etc.) and the private sector, notably private permanent residences used for hosting friends and relations and second homes (defined as 'a permanent building which is the occasional residence of a household that usually lives elsewhere and which is primarily used for recreational purposes') (Shucksmith, 1983, p. 174). Camping and caravanning may constitute an intermediate category wherein private tents or caravans are sited in commercial camping grounds. Certain holiday communities may consist primarily of second homes, and luxury hotels may form the basis of select isolated resorts but most destinations will offer a mix of accommodation types, the mix depending on the nature of the resort and its clientele.

In general, there has been a move away from the traditional serviced type of accommodation provided by hotels and guest-houses to more flexible and functional forms such as the self-contained motel or the rented apartment. Flexibility in ownership is also apparent. Resort apartments or condominiums in many places may be purchased outright, under a variety of lease-back arrangements, or, more recently, on a time-sharing basis whereby a series of owners acquire rights to a property for specified periods of the year. Various collective types of accommodation have also emerged in Europe, particularly in association with policies of social tourism. These usually take the form of holiday camps, where individual bedrooms may be offered along with communal dining halls, lounges and a variety of entertainment or recreational facilities. Many such facilities are sponsored by the State, local authorities or trade unions, although the Club Méditerranée with its pseudo-native villages is essentially of the same format.

Other facilities and services

Besides the provision of these immediate facilities, quite a range of supporting services will be required by the tourist. A variety of shops will be needed, some oriented specifically to the tourist, such as souvenir or sporting goods shops, and others supplying a general range of goods: for example, pharmacies, foodstores or clothing shops. Restaurants, banks, hairdressers and medical centres are amongst the other services needed. Many of these auxiliary services and facilities may serve a predominantly residential clientele. The thresholds for services vary according to the frequency with which they are used. Defert (1966) has proposed a hierarchical model for the development of these services in a traditional resort (Fig. 2.1). Those used every day, such as dairies, cafés and grocers' stores will be the most numerous and amongst the first to be established whereas the higher order services such as jewellers and furriers will come at a later stage when a much larger clientele exists. Today, however, resort development may be so rapid that some luxury services are provided from the outset.

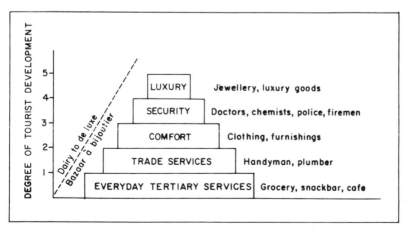

Fig. 2.1   Development of shops and services in a tourist resort
(*Source:* after Defert, 1966).

Infrastructure

An adequate infrastructure will be needed to support the facilities and services outlined above. In addition to the transport infrastructure (roading and parking, airfields, railway lines, harbours) there are the public utilities in the form of electricity and sewage disposal. Much of this infrastructure will also serve the resident population or other needs (e.g. agricultural) but, depending on the type of development, it may also be developed, or expanded, expressly for the tourist. The critical point regarding infrastructure is that although it is essential, it is basically a charge on development. With a few exceptions, such as toll roads, the infrastructure does not itself generate revenue directly. There is little money to be made from disposing of sewage and failure to provide adequate treatment facilities has given rise to one of the more common adverse impacts of tourist development as numerous examples throughout the world testify.

Transport

Historically the development of tourism has been closely associated with advances in transport technology which have facilitated access between markets and destinations. The early development of spas and seaside resorts depended largely on the development of the railways (Gilbert, 1939; Heeley, 1981). In the post-war period, the rapid rise in automobile ownership has been responsible for the vast increase in domestic tourism in Western societies and improved aircraft technology has led to a boom in international travel. However, in addition to increasing the volume of tourist traffic, these advances in transportation have also modified the patterns of tourist flows and hence the patterns of development (Fig. 2.2). Lundgren (1972) and Rajotte (1975) note how travel patterns have become much more flexible and diffuse, with the ubiquitous automobile replacing the linearly constrained railways or river steamers. Studies of the period, such as those by Carlson (1938) and Deasy (1949), also show the effect that the motor car was beginning to have on travel patterns and the rise and fall in the popularity of destinations and types of accommodation. In particular, development was spread beyond those localities served by the traditional means of transport. Large terminus hotels, for example, gave way to highway cabins and eventually to motels. However, Lundgren (Fig. 2.2) also suggests that the subsequent construction of major highway networks has again 'canalized' travellers by the superior accessibility they provide. The modern air-based travel system is nodal in nature, with the expansion of international tourism in particular, being associated with the development of new nodes (airports) and routes linking those nodes (Thurot, 1973). Much of the

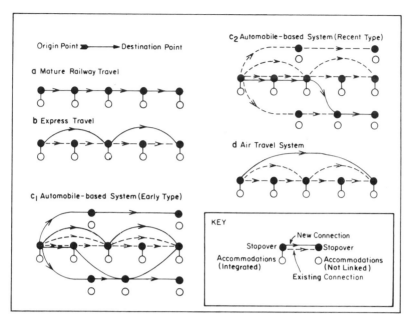

Fig. 2.2   The evolution of tourist travel systems
(*Source:* Lundgren, 1972).

early growth in tourist travel by air involved scheduled services which still remain important especially in regions such as the South Pacific. The development of mass tourism in the Mediterranean, however, is largely due to the emergence of charter airlines which have linked tourist markets and destinations more directly to each other (Pearce, 1987b, 1987c). As aircraft technology has improved, major tour operators have emerged and the relative costs of charter travel have decreased, newer and more distant markets in the Mediterranean have been opened up. The average passenger stage distance of British charter traffic within the Europe–Mediterranean region increased from 1175 km in 1963, to 1381 km in 1969, to 1530 km in 1974 (McDonnell Douglas, 1977) as Spain was joined by such destinations as Greece, Malta and Tunisia.

Transport services within destinations are also important. Some of these may be tourist-oriented (e.g. sightseeing tours, gondolas to lookout points) while various forms of public transport (buses, metros, etc.) used by tourists will primarily serve the local public. Links between the long-haul and local transport systems are also critical (e.g. transit services from airports to downtown or resort hotels). At each scale, these transport systems can be defined in terms of mode, routes and types of operation (e.g. scheduled/non-scheduled).

Market-oriented services

Transport to destinations and the provision there of the facilities and services outlined above may also be complemented by services provided within the market itself. As travel to specific destinations is influenced by the public's awareness of them, many national and regional tourism organizations establish offices in their major markets to promote their country or region and make available information to potential tourists. Outward (and often return) travel is purchased in advance in the market areas, either from the offices of the carriers, especially airlines, or retail travel agents (ITQ, 1984; Holloway, 1985). Both carriers and travel agents may also reserve or sell other items of the tourism experience (accommodation, sightseeing, etc.) either individually or combined as a package or inclusive tour put together by a tour operator. The European Civil Aviation Conference (ECAC, 1981, p. 1) defines an inclusive tour as:

> a round trip or circle tour performed in whole or part by air, organized by a tour organizer and offered to the public at a comprehensive published price including, besides air transport, accommodation for the duration of the trip, surface transport and, where appropriate, other amenities. An inclusive tour is normally paid for before departure, is for a pre-determined period, and is to an announced destination or destinations.

Package tours may also involve ground rather than air transport. Such packaging not only reduces the tourists' need to organize their holidays but also usually brings down the price of travel and for these reasons has been an important factor in the development of mass tourism (Burkart, 1971; Wahab, 1973; Burkart and Medlik, 1974). Tour operators may be based either in the markets (outbound tour operators) or the destinations (inbound tour operators). Other businesses in the market area may also benefit from tourism: banks (currency exchange), insurance companies (travel insurance), shops (cameras and films, sportswear and equipment, travel goods . . .).

Given the range and internal diversity of the different tourism sectors, it is difficult to generalize about the relative size and importance of each sector. Table 2.1 gives some indication of the investment costs at the destination. Although investment in each sector will vary from project to project, accommodation and infrastructure will normally necessitate the heaviest outlays. However, when the market and transport components of the system are also taken into account a different structure emerges. Table 2.2 depicts the average cost structure of an inclusive tour sold through British travel agents. This shows that just under three-quarters of the cost relates to the actual elements of the holiday itself, with those costs being shared almost equally between transport and accommodation. Market-oriented or located costs – marketing, administration, operator's margin and travel agent's commission –

Table 2.1 Capital costs in tourist development

Sector	Average proportions (%)
Accommodation and catering	50–60
Other tourist facilities	10–15
Technical and service infrastructures	15–20
Vocational training promotion, publicity	5–10
Protection and enhancement of resources	5–10

(*Source:* Lawson and Baud-Bovy, 1977, p. 21).

account for the remaining 28 per cent of the package. The latter elements, however, will be non-existent or negligible for other forms of tourism, such as informal domestic or international travel by private car to second homes or the residences of friends and relatives. The travel and stay costs of such tourism will also differ from those shown in Table 2.2.

Successful tourist development depends in large part on maintaining an adequate mix, both within and between these sectors. Natural or historical attractions might be complemented by purpose-built man-made ones to broaden the appeal of the resort or to capitalize on the existing market (*après-ski* entertainment at ski-fields, the lions

Table 2.2 Cost structure of inclusive tours sold through British travel agents

Item		%
Cost of package		
accommodation	36.0	
air transport	34.0	
transfer at destination	0.8	
ground cover (destination staff)	1.2	
Sub-total		72.0
Marketing		3.7
Administration		5.3
Operator's margin		9.0
Travel agent's commission		10.0
Total		100.0

(*Source:* ITQ, 1984).

of Longleat or the wax museums at the Niagara Falls). A range of accommodation at a resort will reduce dependence on a single market. The balance between sectors can be expressed in terms of quality and quantity. Provision of top-grade accommodation at a mediocre attraction, for example, is unlikely to prove viable. Unless certain facilities are to be overcharged or underutilized, the capacity of each of the sectors must be comparable, taking into account, of course, the extent of non-tourist use. In respect to use it is important to note that the tourist product, unlike most others, cannot be stockpiled and resold at a later date. There is no market for last night's unsold hotel bed or yesterday's unoccupied plane seat. Moreover, most tourism plant is fixed in space and cannot be transferred or shipped to new markets if demand changes as is the case with many producer goods. Tour buses and aircraft may be re-routed or re-scheduled but tourist facilities such as hotels cannot be redeployed. Their initial location is thus a very critical factor.

AGENTS OF DEVELOPMENT

The review in the preceding section has shown tourists seek and use a wide variety of services and facilities, with considerable internal diversity being found in the six major sectors outlined. One consequence of this is that the provision of these services and facilities characteristically involves a wide range of agents of development. Some of these will be involved directly and primarily with meeting the needs of tourists, a role that has fallen predominantly to the private sector in most countries. Other agents will facilitate, control, or limit development, often in a general sense, through such means as the provision of basic infrastructure, planning or regulation. Such activities have commonly been the responsibility of the public sector with government at various levels being charged with looking after the public's interest and providing goods and services whose cost cannot readily be attributed directly to groups or individuals. A large number of organizations, particularly international and inter-governmental ones, have also played a role in fostering tourist development in many places.

Private sector

The private sector's prime concern in developing tourism is with making a profit. The nature and extent of private sector participation is influenced by this basic goal and by the multi-faceted and geographically complex nature of tourism already outlined. Attention is frequently drawn to the larger and more visible developers and operators, the multinational hotel chains and airlines, but these are complemented by a multitude of small- and medium-scale operators and businesses, particularly in the attractions and supporting services sectors. Richter (1985a) observes that although there are very large multinational companies such as Holiday Inn and American Express in the United States, 99 per cent of US travel industry businesses are classified as small businesses. Heeley (1986b, p. 76) points out that self-employed and working owners make up 10 per cent of the 2 million or so people working in the UK tourist and leisure industries, 'a statistic which testifies to the numerical dominance of the small independent operator'. At the same time he notes that market dominance and industry leadership lies with the larger companies.

The mix of small- and large-scale units and operators arises primarily out of the different opportunities which exist within and between the tourism sectors discussed earlier. Certain types of activity or facility are less demanding in terms of capital, know-how and entrepreneurial skill than others which are capital-intensive, have a relatively long pay-back period and require high levels of technical and managerial expertise. In the accommodation sector, for instance, camping grounds, bed-and-breakfast establishments and motels lend themselves more readily to individual owner operation than do large hotels. Holiday home development may be characterized by a multitude of individual initiatives in the construction and conversion of detached dwellings or by large-scale condominium development undertaken by a few major real estate developers. Airline operations are usually the preserve of major companies while transportation within destinations is often carried out by small-scale owner-drivers. In developed countries, domestic tourist travel is overwhelmingly by private automobile. Retail travel agencies which sell other companies' products and carry no stock of their own also lend themselves to small-scale operators. In the attractions sector, which is perhaps the most diverse of all, niches exist for a wide range of entrepreneurs from the local hirer of pedalos or jet boat driver to major ski-field and theme park operators.

If the range of opportunities has permitted the participation of small-scale operators, the profit motive, in tourism as in other economic spheres, has encouraged moves to large-scale operations and a concomitant increase in the size of companies involved. Some of this increase has resulted from technological advances. The transition from DC3 aircraft to jumbo jets, for example, not only brought the relative cost of travel down but also contributed to the emergence of large-scale tour operators who could afford and were able to take full advantage of these changes. Much of the growth in large tourism companies arises out of the very competitive nature of many tourist markets and destinations and a drive for increased market share and economies of scale.

Plate 1   Attractions in Queenstown, New Zealand: in this sector 'niches exist for a wide range of entrepreneurs'

Growth in the scale of operations and concentration within the industry has commonly come about through horizontal or vertical integration (Baretje and Defert, 1972; IUOTO, 1975; Wahab, 1975; Holloway, 1985). Horizontal integration involves expansion within a given sector of the industry, for instance through the development of hotel chains or network expansion by the acquisition of the routes of other carriers. Vertical integration, on the other hand, brings together within one organization different sectors of the industry, as for example when an airline acquires hotels or a tour operator opens retail travel outlets. Integration can occur internally by establishing a new product or enterprise from within a company's existing resources or externally, by the acquisition or control of other firms.

As in other industries, expansion through integration brings general technical and financial advantages. Increasing size may mean a firm is able to take full advantage of computerization, an important consideration in tourism as far as reservations are concerned. Or it may mean the larger company can establish a research section which gives it an edge over its competitors. Scale economies can be achieved, for example bulk purchasing or bringing down fixed costs per capita, which can be passed on to customers in the form of cheaper prices. Greater size may also mean more clout in dealing with other operators, for instance in acquiring hotel rooms and seats on aircraft in peak seasons or at cheaper rates. Handling larger volumes of tourists with lower profit margins may also bring down prices which further encourages the development of a mass market.

Horizontal integration means companies can not only achieve some of these general goals but also accommodate more readily some of the geographical dimensions of tourism. Acquiring tourist enterprises in coastal and alpine areas, for instance, may enable the seasonal transfer of staff and a more efficient year-round use of resources. Feeder routes may be acquired to boost the volume of traffic on major scheduled services or, where circuit travel is important, hotel chains may internalize business by channelling guests from one of their hotels to the next. For some segments of the market, the existence of a reputable brand hotel in an exotic destination offers security and familiarity, a base from which the unknown can be explored.

Vertical integration also enables some of the geographical characteristics of tourism to be overcome while at the same time 'closing the circle' of services and activities which make up the tourism experience. Many early coastal resorts were developed by railway companies seeking to generate new business. The Canadian Pacific Railway was a major hotel developer and railway companies have also been important in developing tourism in Japan (Yamamura, 1970). Later, international airlines acquired hotel subsidiaries (Pan American and Intercontinental Hotels, Trans World Airlines and Hilton Hotels, JAL and Nikko, Air

France and Meridien). Their motives for doing so vary considerably (Centre on Transnational Corporations, 1982, p. 36):

> Sometimes this involvement is induced by the need to provide accommodation for airline passengers, or to establish links with a destination in order to obtain traffic rights. In other cases, it may be prompted by a desire to exercise control over passenger movements and to pre-empt a profitable market for hotel accommodation from indigenous investors. Alternatively, the involvement may be more in the nature of portfolio investment, reflecting a desire to enter a profitable and rapidly expanding market.

Some of the arrangements noted here have not survived (Pan American sold Intercontinental to Grand Metropolitan after deregulation and Ladbrokes, the British property and leisure industry conglomerate, acquired Hilton in 1987 from Allegis, née United Airlines) but others continue to have close links. Almost half of the hotel bookings made through JAL in 1985 were routed into the airline's chain of 120 hotel affiliates in thirty-five countries. Other airlines say that passengers stay at affiliated hotels only when the quality and price suit them (*Economist*, 5 September 1987).

Vertical integration has been especially evident in the package tour industry (Pearce, 1987b). Major tour operators have sought to keep prices down, ensure access to seats and beds, and maintain some control over quality by progressively becoming involved in all sectors and stages from the market (retail outlets), through the transport sector (control of an airline) to the destination (ownership or control of hotels). Initially many tour packages in Europe were put together by retail travel agents who were able to charter small aircraft and book a corresponding number of hotel rooms. As competition for the ITC market increased, profit margins became progressively smaller and survival depended on handling a large volume of traffic (Wahab, 1973). The increase in aircraft size further encouraged the trend towards larger scale operations and several major tour wholesalers began to emerge. By the late 1970s, the top six tour operators handled over 75 per cent of the ITC demand within the United Kingdom, with the largest, Thomson, handling over 1 million passengers a year (Guitart, 1982; Heape, 1983). Several of the major charter airlines have been acquired by tour operators. Some tour operators have also acquired interests in the hotel industry in the destination countries (Gaviria *et al.*, 1974; Centre on Transnational Corporations, 1982). Increasing size and integration have allowed some companies to become very competitive, with their high-volume low-tariff strategies striking a responsive chord among West European holidaymakers. Size alone, however, has not prevented tour operators from failing.

Tourist activities and firms may also be acquired by companies or conglomerates whose prime centres of interest lie in other sectors of the economy (Centre on Transnational Corporations, 1982). Expansion into tourism may represent diversification into what has been a growth sector in many places. Or it may represent another form of integration, as when brewing/drink combines acquire hotels because of the guaranteed sales outlets they represent (Heeley, 1986b). Capital injection from non-tourism sources may be welcomed given the long payback period of many tourism investments. Association with a more broadly based enterprise may also ease cash flow problems of seasonally constrained tourism firms. Hotel or condominium development may prove attractive to property developers, not only for the longer term real estate gain but also for the increased public image that such glamorous holdings can bring. As the *Economist* (5 September 1987, p. 68) commented in the lead up to the sale of Hilton Hotels:

> At a price tag of $1 billion, Hilton International would be yielding a pre-tax return of less than 10% even if the chain hits its target of 15% annual growth in pre-tax profits for the next five years. Eurobonds would provide a much better return over this term. But people do not always buy hotels on such dry calculations. Eurobonds do not have discos in their basements.

Integration may continue to a point where corporations extend their activities to a range of countries so that their operations become multinational or transnational in nature. The most objective and comprehensive study of transnational involvement in international tourism is that by the Centre on Transnational Corporations (1982). For the purposes of that study, transnational corporations (TNCs) were defined (p. 2) as 'not only foreign firms with direct investments in a particular host country but those firms having all major forms of contractual arrangements, and enterprises in host countries'. Three major types of TNCs were identified and examined: hotels, airlines and tour operators. As the latter two have been discussed earlier, hotels are dealt with in more detail here.

A general overview of the distribution of TNC-associated hotels abroad is provided by Table 2.3 which shows two-thirds of the hotels were independent of airlines while just over a quarter were associated with them. In 1978 the hotels were distributed almost evenly between those in developed market economies, particularly Canada and Europe, and those in developing economies, notably Asia. United States based hotel chains accounted for 56 per cent of the rooms abroad, French chains 13 per cent and British ones 12 per cent. Chains based in developing countries had only 4 per cent of the total rooms held by TNCs although significant domestic chains are also found in many of these countries (e.g. El Presidente, Mexico; Merlin, Malaysia). Considerable variation occurred in the relative importance of TNC hotels in host countries (0.6 per cent, USA; 67.2 per cent Martinique). With the exception of New Zealand, all ratios of 10 per cent or more

Table 2.3 Distribution of transnational-associated hotels abroad, by main activity of parent group, 1978

Parent group	Number of transnational corporations	Trans-national-associated hotels abroad		In developed market economies		In developing economies	
		Number	%	Number	%	Number	%
Hotel chains associated with airlines	16	277	27.1	113	21.0	164	33.7
Hotel chains independent of airlines	56	687	67.0	384	71.2	303	62.4
Hotel development and management consultants	3	15	1.5	1	0.2	14	2.9
Tour operators and travel agents	6	46	4.5	41	7.6	5	1.0
Total	81	1025	100.0	539	100.0	486	100.0

(*Source*: Centre on Transnational Corporations, 1982).

were in developing countries or territories. In general it was found (p. 20) that:

> the larger the absolute size of the hotel industry in a particular country, the lower the foreign participation is likely to be. Conversely, the foreign participation ratio is likely to be higher in countries in which hotels are concentrated in central cities than those in which they are concentrated in resort areas, because transnational corporations are more likely to be involved in hotels that service business rather than those that cater for holiday tourism.

In terms of motivation, the Centre on Transnational Corporations concluded (p. 47):

> The basic reason for the extent of transnationals in hotel operations abroad is that the benefits generated by the transnational corporation as a whole is greater than the sum of the separate parts. This in turn, is partly a matter of the size, geographical diversification and business strategy of transnational corporations and of the economies of scale associated with

advertising, the provision of first-rate training facilities, good promotional prospects for higher management and technical staff, bulk purchasing and marketing (including computerized global reservation systems) etc. On the other hand, the Centre also observes that:

> Domestic hotels are often more flexible in their operations than the centrally controlled transnational corporations and are, therefore, potentially more capable of adapting to local conditions. This may result in lower operating costs, greater diversity in the range of services offered by the hotel . . . and greater ability to comply with the economic, social and cultural objectives of the economy, for example by processing imports from local subcontractors, designing the hotel in order to maximize the use of local materials and reflect indigenous architecture, employing indigenous managements etc., and thereby minimizing the transfer of earnings abroad.

These points are explored further in subsequent chapters.

More generally, there are limits to integration, particularly in a service industry such as tourism (Baretje and Defert, 1972). Better personal service is not necessarily guaranteed by increasing the scale of operations, rather it may be lost. Standardization of hotels may have provided a certain level and type of service and package tours may have brought overseas travel within the reach of many who had never holidayed abroad previously, but the demand for tourism is not so uniform that such strategies will appeal to everyone. On the contrary, diversity and personalized holidays may become increasingly important and the small and medium enterprises which contribute to this variety and may be better able to offer these services do not seem destined to disappear. The inherent diversity of tourist attractions is also a reason why this sector has been less subject to big company involvement (Heeley, 1986b).

As well as this direct involvement in tourism operations, the private sector may also provide many of the technical inputs into the development process (market research, architectural design, construction, etc.) and provide a large share of the capital for investment in tourism (Mill and Morrison, 1985).

While profit-making is the driving force for this group there are also ventures which appear not to be characterized by a sound economic rationale but influenced more by the whims of a managing director or the efforts of an individual who is attracted by the apparent glamour of the tourist industry, by the desire to create something or merely because of an enthusiasm for skiing. The individual second-home owner will be motivated as much by the desire to fulfil his own leisure ambitions as to develop or acquire an investment. Private, non-profit-making organizations may also be indirectly involved in tourism, for example cultural and historical societies responsible for art galleries, museums or the conservation of historic sites and buildings (Heeley, 1986b).

The public sector

Central or federal government in most countries is complemented by smaller territorial local authorities (boroughs, counties). An intermediate tier of government also exists in many countries, such as the State and provincial governments in North America and the *départments* in France. Moreover, central government consists of a number of different agencies or departments, many of which may be involved in a tourist development programme. Their various activities may be co-ordinated by some national tourism organization. The participation of these different levels of government will depend on the nature and scale of the project but it would not be uncommon for all three levels to be represented in some way in a particular programme.

The public sector may become involved in tourism, directly or indirectly, for a wide variety of reasons.

Economic

Various economic factors may induce the public sector to foster tourist development:

1. Improved balance of payments situation.
2. Regional development.
3. Diversification of the economy.
4. Increased income levels.
5. Increased State revenue (taxes).
6. New employment opportunities.
7. Stimulation of non-tourism investment.

As was noted in Ch. 1, tourism is sometimes perceived, not always correctly, to have a 'special function' and offer certain advantages in these areas, for example, large foreign exchange earnings and employment generation from a limited level of investment (Krapf, 1961; Kassé, 1973).

Social/Cultural

Social considerations are also important. Hughes (1984, pp. 16–17) sees these in terms of 'social benefits' ('the enjoyment and relaxation offered by holidays of whatever kind may in some way contribute to a more stable, civilized and cultured society as well as a more stable one') and 'merit wants' (citizens should not, because of low income or ignorance, be prevented from consuming goods or services which are important to their well-being). Such considerations have given rise to a policy of 'social tourism' in many European countries (Lanquar and Raynouard, 1978; Haulot, 1981). The State may also have a general responsibility to protect the social and economic well-being of the individual, for example through health regulations and consumer legislation or by minimizing adverse social impacts of tourism and other developments. Wood (1984) suggests that the State, especially in multicultural societies, plays an important role as marketer of cultural meanings and as arbiter of cultural practices.

Environmental

The responsibility of protecting and conserving the environment, both physical and cultural, usually falls to the public sector although it might be argued that there is an individual or corporate responsibility here as well.

Political

As international tourism involves the movement of people from country to country, governments may encourage the development of tourism to further political objectives. Cals (1974) suggests that the Spanish government encouraged tourist development amongst other things to broaden the political acceptance of Franco's regime. In Israel, the development of tourism has done much to stimulate political sympathy for that nation and to boost national morale (Stock, 1977). In the Philippines, Marcos actively exploited tourism to meet the political needs of his New Society: 'Legitimacy, international influence, foreign economic investment, patronage and personal fortunes represent high political stakes which the administration has garnered through an adroit use of government credit, imagination and Filipino hospitality' (Richter, 1980, p. 238). Richter also notes that: 'Other nations, particularly Korea and Taiwan, are also finding that the political benefits of world travel may be as rewarding as its much vaunted economic advantages.' Elsewhere she observes that 'domestic tourism may also serve important political and cultural goals such as national integration and a sense of national pride' (Richter and Richter, 1985, p. 208). Socialist governments may also find certain forms of tourist development consistent with the pursuit of their ideologies (Hall, 1984).

Public sector involvement in tourist development then usually comes about as part of a broader policy or programme of intervention. Tourism is encouraged as a means of increasing the inflow of foreign currency or pursued for political purposes rather than as an end in itself. The extent of public intervention varies from country to country and is to a large degree determined by the general philosophy and policies of the governments in question. Public sector involvement in tourism is likely to be less in Western liberal economies characterized by free market philosophies than in the centrally planned economies of the Soviet bloc or many developing countries.

41

The general size and characteristics of tourism also influence the degree of public sector intervention. In many developed economies, emphasis has traditionally been given to agriculture and manufacturing, industries which have often attracted large measures of government support which may not be available to service industries such as tourism (Hughes, 1984). Richter (1985a, p. 168) notes that 'while service industries are becoming increasingly dominant in the USA and are a growth area of employment and urban revenue, such industries have not captured the attention of government, business and even academic research that product-oriented industries enjoy. The government is still preoccupied with saving obsolete industries that can never be competitive again . . .'. Richter then outlines a number of features of US tourism which encourage its 'political neglect':

–   those employed in tourism (mainly women and minorities) do not constitute a salient political constituency,
–   tourism is politically unimportant and is perceived as over-whelmingly domestic in character,
–   the tourist industry is fragmented and lacks political muscle.

Not all developed Western economies share the American experience, however. The government of France, for example, has a long history of active intervention in tourism (Jocard, 1965), including direct participation in the development of the coast of Languedoc–Roussillon, one of the largest tourism projects in the world (Pearce, 1983).

Jenkins (1980, p. 27) argues a case for the governments of developing countries to 'positively intervene' in tourism:

When funds are required to support investment in tourism, the government is often the only agency able to raise or guarantee the loan. At the macro level, the government has ultimate responsibility for the allocations of funds and resources for specific sectors. It is a government responsibility to decide on regulations and loans which can affect tourism . . . .

The most direct and explicit manifestation of central government participation in tourist development is usually the national tourism administration (NTA). NTA is used by the World Tourism Organization (WTO, 1979, p. 11) to designate 'the authorities in the central state administration, or other official organization, in charge of tourism development at the national level' (elsewhere these may be known as NTOs (national tourism organizations) or GTOs (government tourist offices)). A 1978 survey by the WTO of 100 NTAs showed two-thirds took the form of ministries or constituted part of a government department. The remaining third had their own legal personality, for example they were government corporations, but were linked to or were under the supervision of the central administration. Considerable variation was found in the responsibilities and activities of the NTAs surveyed but the following were identified as the most common:

– tourist promotion and information,
– research, statistics and planning,
– inventory of tourist resources and measures for their protection,
– development of tourist facilities,
– manpower development,
– regulation of tourist enterprises and professions,
– facilitation of travel,
– international co-operation in tourism.

While the activities of the NTAs may influence the path of tourist development directly, other more general or more indirect powers exercised by other central government agencies may also have a significant and perhaps greater impact. In particular the State has the power to regulate numerous areas which impinge on tourist demand and development. Civil aviation regulations, concerning both scheduled and non-scheduled (charter) operations, can play a major part in determining accessibility and the level and pattern of tourist demand (Kissling, 1980; Jenkins and Henry 1982; Pearce 1987b). Similarly, customs and immigration regulations can foster or impede international tourist movements. Liquor licensing and shop trading laws may also regulate patterns of tourist activity. Conservation legislation can protect a nation's natural and historical heritage. Such measures may run counter to the short-term exploitation of these resources although contributing to the nation's and the tourist industry's well-being in the long run.

Government fiscal policies may also encourage or hinder the development of tourism. Adjustment of exchange rates, both in markets and destinations, may restrain or promote tourist travel with countries such as Spain seeking to maintain a competitive advantage through frequent devaluation of their currencies. Governments may also set legal limits on the extent of foreign investment and the repatriation of earnings and through grants, incentives and taxes can encourage or limit investment in tourism or channel it to certain localities. In the early 1980s, for example, the Turkish government took a series of steps to foster tourism which shifted their emphasis from promotion to the encouragement of foreign investment (Liu, Var and Timur, 1984). Incentives offered to foreign investors included:

– long-term subsidized interest rates,
– five-year exemption from construction and property taxes,
– provision of State land for tourism projects for 99 years at low rental,
– State provision of basic infrastructure in tourist development areas.

Measures were also taken to encourage tour operators through subsidies for charter flights with empty seats and to devalue the lira.

The State may also play a key role as landowner or resource manager, particularly in New World countries where areas of special interest to tourists and tourist developers, for example coastal and upland areas, are in public ownership. Such is the case in the Colorado Rockies where the public domain, controlled predominantly by federal agencies, often begins at the tree-line and extends to the peaks (Thompson, 1971) and in the Southern Alps of New Zealand where most of the high country is Crown land (Pearce, 1985a). In many instances management of these lands is the responsibility of agencies whose dominant function is forestry, soil conservation or some other non-tourism activity and whose policies and attitudes may or may not facilitate tourist development. Management of national parks, which are major tourist attractions in countries such as the USA, Canada and New Zealand, involves maintaining a balance between nature conservation and recreational use.

Responsibility for land in the public domain may also lie with lower levels of governments, such as regions, counties and municipalities. Regional or State governments may also have the power to legislate and territorial local authorities will be able to pass by-laws and introduce building codes and zoning regulations. In terms of physical development, the ability of the territorial local authority to issue or withhold building permits may be its most important function. Provision of much of the infrastructure, the presence or absence of which may facilitate or hinder development, will also often be the responsibility of local and regional government, though larger scale projects may require support from the central administration. Promotion and planning is also carried out at these scales (Richter, 1985b; Smyth, 1986). Indeed Murphy (1985) argues that: 'the opportunities presented by the [tourist] industry and the difficulties arising from its rapid development can best be examined and resolved through a community approach'.

The 'public sector' then is by no means a single entity with clear-cut responsibilities and well-defined policies for tourist development. Rather, the public sector becomes involved in tourism for a wide range of reasons in a variety of ways at different levels and through many agencies and institutions whose interest in tourism is frequently peripheral to some broader area of responsibility. Consequently, there is often a significant lack of co-ordination, unnecessary competition, duplication of effort in some areas and neglect in others.

In Northern Ireland, for example, the major bodies involved in tourism have been the Department of Economic Development, the Northern Ireland Tourist Board and the local authorities with secondary responsibilities lying with such diverse agencies as the Department of Education and the Forestry Division of the Department of Agriculture.

Smyth (1986, p. 126) concludes that without an effective organizational framework there has been a failure there to 'provide those statutory agencies having an involvement in tourism with the direction necessary to enable them to evolve purposeful strategies'. Similarly in Scotland, Heeley (1986a, pp. 68–70) noted 'an absence of clear and unambiguous divisions of responsibilities' and called for a 'clear governmental statement of the three development bodies (Scottish Tourist Board, Highlands and Islands Development Board and Scottish Development Agency), and a co-ordination of governmental activity across the arts, heritage and tourism spheres'. Some restructuring of the administration of tourism in Great Britain has occurred recently with the responsibility for tourism being shifted away from the Department of Trade and Industry to the Department of Employment and proposals being put forward for a new British Tourist Board to take over the functions of the English, Scottish and Welsh Tourist Boards (Heeley, 1986a).

Organizations

Numerous organizations at different levels may also play a major role in the development of tourism. These may be essentially inter-governmental in nature or bring together public and private sector voices and interests. Some will be concerned with development in general, others will have tourism as their sole or prime focus.

International organizations and agencies have played a significant role in promoting international tourism as a means of economic development, particularly in developing countries (Krapf, 1961; Hector, 1978; Wood, 1979; Cazes, 1980a). As Wood (p. 276) points out: 'Underdeveloped countries did not simply stumble onto tourism as a promising way to earn foreign exchange . . . . Beginning with the Checchi Report in the late 1950s, the governments of these countries have been presented with a steady stream of advice and assistance for expanding international tourism.' International tourism was seen at a comparatively early stage to have a 'special function' (Krapf, 1961) and one which was seized upon and has been actively promoted by such agencies and organizations as the World Bank, the Inter-American Development Bank, the United Nations Development Programme (UNDP), the Organization of American States (OAS) and, more recently, the European Community.

International tourism in developing countries has been encouraged in three main ways:

- technical assistance, especially in the preparation of tourist development plans,
- loans for major infrastructural projects,

–   loans and equity investments in privately owned tourism plant, particularly hotels.

Technical assistance and plan preparation have been emphasized by inter-governmental agencies with the OAS, for example, preparing a number of tourism plans for its member states (OAS, 1978a, 1978b, 1978c). A 10-year plan for tourism was one of a number of specific projects making up the Belize Public Investment Project which was initiated by the UNDP as Belize approached independence (Pearce, 1984). Aid agencies have identified tourism infrastructure and plant as appropriate projects for loans and investment. Table 2.4 shows the extent and distribution of tourism commitments by the World Bank Group through to 1977 (Wood, 1979). The World Bank (IBRD) itself and its soft-loan subsidiary, the International Development Association (IDA) have concentrated on tourism infrastructure, with some support going towards the development of tourism plant and providing lines of credit for financing hotel investment (Davis and Simmons, 1982). The International Finance Corporation (IFC), on the other hand, promotes private sector ownership. Three-quarters of the IFC's investment in tourism was to hotels with management or significant ownership by eight major multinational companies such as Intercontinental Hotels or Holiday Inns. Under the Lomé III Convention, the European Community has specifically identified tourism as a means of assistance for the Africa, Caribbean and Pacific (ACP) States via grants and loans from the European Development Fund and the European Investment Bank (Lee, 1987). The European Community also encourages regional co-operation in developing and marketing tourism in the small States of the Caribbean, the Pacific and the Indian Ocean.

But as well as promoting international tourism as a path towards economic development, these agencies have, in the main, also encouraged a particular form of tourist development, one based on large-scale

Table 2.4 Tourism commitments of the World Bank Group through to 1977 by region ($ millions)

Region	IBRD	IDA	IFC	Total
Africa	100.9	14.0	25.3	140.2
Latin America/Caribbean	85.0	—	14.8	99.8
Middle East	—	6.0	—	6.0
Europe	36.0	—	0.6	36.6
Asia	25.0	20.2	17.6	62.8
Total	246.9	40.2	58.3	345.4

(*Source*: Wood, 1979).

projects involving heavy investment in infrastructure and plant and a high degree of non-local participation (Hector, 1978; Noronha, 1979; Wood, 1979). Such projects have not always returned the economic benefits promised nor led to greater self-reliance. Appreciation of these points has not always come quickly to some of the agencies concerned though it must be said that the World Bank and UNESCO jointly funded the first major reviews of the social and cultural impacts of tourism (Thurot *et al.*, 1976; de Kadt, 1979).

Not all tourism activity by general inter-governmental organizations, however, has been directed at developing countries. The Organization for Economic Cooperation and Development (OECD), for instance, followed its early seminar on tourism and economic growth for its developing and member countries (OECD, 1967) by a major research and policy programme on tourism and the environment (OECD, 1981a). The OECD also regularly publishes policy reviews and statistical information which enables the development of tourism in its member countries to be monitored.

In addition to targeting tourism in its co-operation programmes, the European Community is also evolving Community tourism policies (Airey, 1983; Pearce, 1988a). Recent reports and documents reflect a more explicit and enthusiastic recognition of tourism within the Community, with its role for employment creation being complemented by its importance for European integration. Objectives of Community action in this field include: facilitating tourism within the Community, improving the seasonal and geographical distribution of tourism, making better use of Community financial instruments, providing better information and protection for tourists, improving working conditions in the tourist industry and increasing awareness of the problems of tourism and organizing consultation and co-operation. So far, most action has occurred through the European Regional Development Fund (Pearce, 1988a).

The largest specifically tourism-oriented inter-governmental organization is the World Tourism Organization (WTO). Formed in 1975 to replace the former International Union of Official Travel Organizations (IUOTO), the WTO is based in Madrid and has a membership of over 100 official government tourism agencies together with a large number of affiliate members representing other tourism-related organizations. The best known of the WTO's activities is its annual publication of tourist statistics which enable trends in the growth of tourism to be monitored. Compilation and diffusion of these statistics is complemented by an active research and technical assistance programme (Gee, 1984). In addition to specific missions, the results of these activities are presented at seminars and conferences and through an extensive range of publications which help raise the level of technical input into tourist development and also influence attitudes towards tourism. The WTO also has a general goal of facilitating tourism throughout the world and its Manila Declaration, resulting from the World Tourism Conference of 1980, is a major statement on global tourism issues (Gunn and Jafari, 1980; WTO, 1980a).

Other international tourism organizations have a more regional focus, such as the European Travel Commission which brings together the national tourism organizations of Europe 'to increase the level of tourism from other parts of the world to Europe as a result of its marketing activities' (O'Driscoll, 1985). Marketing is also a major thrust of the Pacific Asia Travel Association (PATA), an organization with private sector members, particularly major tourism companies, as well as NTAs. In particular, PATA undertakes collective market research programmes in major markets which are beyond the scope of individual members. PATA also provides technical assistance for tourist development, notably through missions to individual countries or regions.

A wide range of national, regional and local tourist organizations is also to be found in most countries. Many of these are sectoral and serve to further the interests of a particular group, for example hoteliers or tour operators. Others are umbrella organizations, for example the New Zealand Tourist Industry Federation, which draw together representatives from different sectors enabling them to speak with a stronger united voice and to co-ordinate activities such as marketing, promotion, planning and research. Such co-ordination is particularly important given the multi-faceted and fragmented nature of the tourist industry and the diverse roles and functions of the public and private sectors. Effective input into the formation of public policy regarding tourism, for example, is difficult without a forum for considered and coherent views to be developed and expressed.

Public and private participation

Given the diversity of the elements to be supplied, most tourist development will involve a mix of these different agents of development. These may come together in a variety of ways, informal or structured, competitive or co-operative. The initiative may come from either the public or the private sector. Government at any level may solicit private investment and development through the provision of infrastructure, a development plan and fiscal incentives. Or, local authorities may have to extend their public utilities or modify their town plan in response to pressure from private entrepreneurs.

Certain common roles and responsibilities can be recognized.

Because of the scale of development and the element of common good, provision of infrastructure is a widely accepted task of public authorities. Exceptions occur nevertheless, as in New Zealand where most ski-field access roads have been built and are maintained by private developers. Likewise, isolated resorts commonly provide their own water supply and sewage-treatment plants. International promotion and marketing is also often seen as providing a common good and for this reason has been one of the major functions of NTAs. Nevertheless the private sector, particularly international carriers, also invests large sums in these activities. Preservation, conservation and enhancement of natural and to a lesser extent historical attractions are a major concern of the public sector, expressed most notably in national parks. Although involved in all sectors, private enterprise has been most active in the accommodation and supporting facilities sectors and in providing purpose-built attractions. Government-owned hotel chains, such as the Tourist Hotel Corporation in New Zealand and the Paradores in Spain, also exemplify State participation in these sectors.

The mix of private, public and possibly organizational involvement will depend on a number of factors including the scale and nature of the project, the stage of development and general government policies regarding free enterprise and State intervention. According to Cazes (1978), we have gone from the nineteenth century situation characterized by a loose grouping of individual actions to a more systematic, quasi-institutionalized organization in which central government formulates the programme and develops the basic infrastructure, leaving private enterprise to provide the tourism plant and associated services. However, this is perhaps truer of parts of Europe and certain developing countries than North America and other Western societies where less formal, unstructured, competitive development tends to be the norm. Moreover, as has been noted earlier with respect to Scotland and Northern Ireland, the roles of different public agencies are not always neatly delineated and co-ordinated.

A more formal structuring of responsibilities and co-ordination of the different agents of development frequently comes as the scale of the project increases. Such has been the case with the development of the Languedoc–Roussillon project, which was begun in the early 1960s and encompasses the 180 km of the Mediterranean coastline stretching from the delta of the Rhône to the Spanish border. A clear and co-ordinated division of responsibility was required to undertake such a large-scale programme. No one organization had the resources or competence to take everything in hand yet the success of the operation depended heavily on all facets of development being carefully co-ordinated. The roles of the State, the local and regional authorities and the private sector were therefore defined right from the outset (Fig. 2.3).

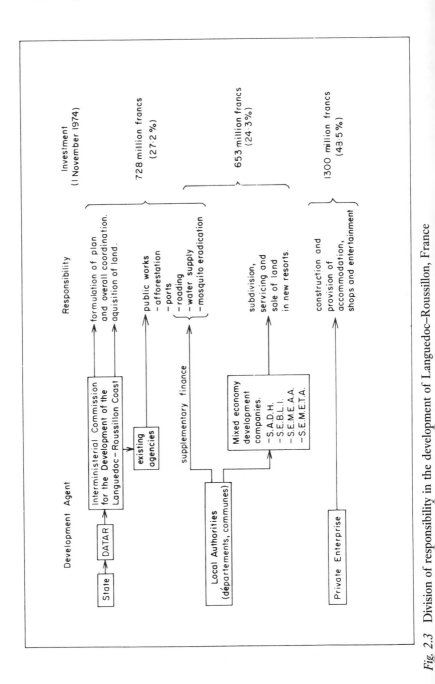

Fig. 2.3 Division of responsibility in the development of Languedoc–Roussillon, France

The State

The State was firstly responsible for drawing up the overall development plan. Following this plan it oversees and controls all subsequent phases of the operation through a small but important study team. Secondly, as was noted above, the State acquired all the necessary land for the operation to go ahead as planned. Thirdly, it was responsible for undertaking major infrastructural works; the road network, the ports, afforestation, water supply and much of the mosquito eradication. Finally, State finance was made available to the other authorities to undertake their responsibilities.

Administrative innovations facilitated the work of the State. In 1963 a Mission Interministérialle was created under DATAR (Délégation à l'Aménagement du Territoire et à l'Action Régionale), drawing together representatives of the interested government ministeries (finance, interior, development, agriculture and tourism). The Mission was a surprisingly small organization, much of its work being the responsibility of a small study team, consisting of a dozen people. Their role was very much a co-ordinating one, many of the initial studies and most of the actual development work being undertaken by the appropriate government department to which is transferred the necessary finance from the Missions's central budget.

Local and regional authorities

The territorial authorities provided supplementary finance for the regional works. They also participated directly in the development of the new resorts through the creation of four mixed-economy companies. The SADH (Société d'Aménagement de l'Hérault), for example, which is responsible for the development of La Grande Motte and Carnon, has as its major shareholders the Département de l'Hérault and three local municipalities. Other shares are held by different government agencies: SCET (Société Centrale pour l'Equipement Touristique); Compagnie Nationale d'Aménagement du Bas-Rhône Languedoc (CNABRL) and various Chambers of Commerce. Each of the four mixed-economy companies is responsible for servicing and developing the land acquired from the State for the major new resorts. Having subdivided it and installed the necessary services – water, electricity, sewerage, roads, parking areas, telephones – these companies then resell the land at a price to cover initial costs and these improvements to the private sector.

Private sector

Thus it was only when the plan had been established and the necessary infrastructure undertaken that the private sector entered the development process to construct the hotels, apartments, villas, camping grounds, shops, entertainment and other facilities. Each particular project had to conform to the established plan and have the accord of the appropriate chief architect but prices and the decision to sell or let were determined by the individual promoter. Regional firms have played a major role here, particularly in the construction and sale of apartments. Some 60 per cent of apartment construction at La Grande Motte and Carnon and virtually all such development at Port Carmargue has been by promoters from the three local departments of the Gard, Hérault and the Aude. British, German and Japanese capital has been invested in new hotels, a sector which has attracted little interest from the French.

From a technical point of view the Languedoc–Roussillon operation has been reasonably successful. The unity of direction, the division of responsibility and the comprehensive nature of the operation have brought results. Development has been reasonably orderly and coherent but the failure formally to designate full responsibility for marketing, promotion and sales contributed to a slower than planned rate of growth.

Clearly many other forms of tourist development are possible as the different agents of development combine or compete in various ways to deliver the elements outlined earlier. Such agents, however, do not act in a vacuum and an appreciation of the context in which tourism develops is critical for understanding the process of tourist development which occurs in any place and its resultant impacts.

THE CONTEXT OF DEVELOPMENT

As was noted in Ch. 1, tourism has developed in a number of different physical, social/cultural, political and economic contexts. Consideration of contextual characteristics – the nature of the place in which tourism develops – is important, for the context will influence the way in which tourism evolves and will condition the impact which tourism will have. These impacts will in turn modify the places concerned and contribute to a redefinition of their character and geography.

The physical backdrop can influence tourist development in several ways. Most importantly, physical attributes such as climate and landscape may have a major bearing on the attractiveness of an area which will influence the type of tourism which develops. Physical characteristics may also facilitate or limit accessibility by different modes of transport. Many of the environments which appeal to tourists – alpine areas, coasts and tropical islands – may be especially dynamic or fragile areas whose balance is readily disturbed by tourist and other developments. As Pearce and Kirk (1986, p. 3) point out:

> Coasts, particularly beaches, are amongst the least stable and most physically changeable of the Earth's landform systems. Beaches exist because supplies

of unconsolidated sediments are delivered to the margins of water bodies, and because waves and currents can redistribute and shape the accumulations. Any factor, natural or man-originated, which alters the supply, throughput and loss of sediments and/or the distributions of wave energy and currents in time and space, will have its repercussions on the form and position of the shoreline.

Similarly, the social and cultural characteristics of a host society will influence its attractiveness to tourists, the process of development and the nature and extent of the impacts which occur. A distinctive culture may appeal to certain groups of travellers. Many possibilities exist: remnants of past cultures (Mayan temples, Egyptian pyramids), distant, exotic cultures (Polynesian folk dances), or distinctive, geographically confined societies (the Amish of Lancaster county, Buck, 1977). Class and political structure may also determine the type of tourist development that occurs. Wood (1979, pp. 282–3) argues that one of the reasons for multinational involvement in luxury hotel development in South East Asian societies is that 'joint enterprises with multinationals may allow politically dominant classes without a strong economic base to circumvent their economically dominant class (and ethnic) competitors. This is likely to be a factor in countries like Malaysia, Indonesia, and Thailand, where alliance with multinationals avoids the necessity of doing business with the Chinese commercial classes'. In terms of impacts, Maurer (1980) suggests:

> It would be fallacious to compare, as it is often done, some of the Indonesian microcultures particularly affected by tourism penetration, which have developed for historico-cultural reasons a relatively strong capacity of resistance and are located in a rather protective geographical environment, with the fragile and faltering cultures of the tiny exposed and unprotectable Hawaiian or Bahamas islands.

In general, more attention has been paid in the literature to the economic context. Particular emphasis has been given to the economic characteristics of developing countries and how these have influenced the decision to develop tourism and conditioned the effects of that development (e.g. Bryden, 1973; Kassé, 1973; Goonatilake, 1978; Britton, 1982). However, in the tourist literature there have been few attempts to analyse systematically differences between developed and developing countries, let alone socialist economies, nor to take account of internal variations within any one group. One notable exception is Young's (1977) study in which she examines the interrelationships between three types of tourism – comprehensive, luxury and plantation – and three sets of structures – ecological, economic and political – throughout the Caribbean.

The different factors outlined here combine to produce an overall context in which tourist development occurs. This can be seen in the case of tourism on small islands where the combination of smallness and insularity produces spatial structures and development processes which are more evident there than in most mainland countries and destinations (Pearce, 1987a, pp. 151–65). Tourist accommodation in islands tends to be concentrated in a number of coastal localities in or adjacent to the major urban centre and close to the international airport. The predominance of sun–sand–sea tourism, especially on tropical and subtropical islands, is a direct consequence of an insular situation and the limited range of other possible tourist resources. The absence of relatively low-cost land access reduces the range of possible markets and the types of development which may occur. This pattern is reinforced by the limited local demand which stems from the size of the islands' own markets and the small distances involved. The emphasis on air travel and structured holidays for foreign tourists not only leads to the concentration of hotels close to the airport and major urban centre but increases the likelihood that the island tourist industry may be controlled by external developers. The effect of these factors on small islands may also be reinforced by relatively low levels of economic development which in turn result in varying degrees from the conditions of smallness and insularity as well as from the broader political–economic context (Hills and Lundgren, 1977; Britton, 1982).

In terms of the environmental impact of tourist development, McEachern and Towle (1974) underline the fragility of island ecosystems. They note that small islands are seldom well-endowed with resources and consequently are liable to over-exploitation. The salination of fresh water supplies on Mallorca is a case in point. Moreover, isolation and circumscribed space tend to produce specialized floral and faunal communities which tend to be less tolerant to environmental changes, particularly those resulting from exogenous pressures.

Contextural factors can also give rise to considerable diversity within island groups as Young (1977, p. 665) showed in her analysis of the Caribbean: 'Comprehensive tourism is found on islands with less rainfall that makes such islands attractive in general to tourists. . . . Luxury tourism tends to be found on relatively flat islands with a good system of all-weather roads. Plantation tourism tends to be found on the larger agricultural islands.' Young also found plantation tourism to be positively related to repressive government. In the Balearic Islands, Bisson (1986) cautions against explaining the dominance of Mallorca solely in terms of the larger size and resource base of the island and the earlier development there of international air connections. Rather he points to other less obvious factors such as the underlying agricultural and tenurial structures. In the early 1960s when charter tourism to Mallorca 'took off', much of the coast in the commune of Calvia had fallen into the hands of financial and property development companies as a result of the indebtedness which had arisen out of the break-up of previously large estates amongst a number of heirs. Such a situation lent itself readily to the urbanization of the coast. In contrast, tourist development on Menorca was delayed by a more conservative land-holding system with succession there being to a sole heir and share cropping prevailing on the large estates. Industry was also relatively more important on Menorca, further delaying diversification into tourism.

CONCLUSIONS

Tourist development involves the provision of a wide range of services and facilities which are furnished, directly or indirectly, by a variety of public and private sector agents characterized by their diverse interests, roles and responsibilities. This composite and multi-faceted nature of tourist development is perhaps its most dominant feature and one which renders difficult the delineation of other general characteristics. Certain common dimensions of public and private sector involvement in tourist development have been identified, such as the private sector's concern with profit-making and the broader responsibilities of public authorities, a division of interest which is reflected in the former's dominance in the accommodation and purpose-built attractions sector and the latter's role in providing infrastructure and conserving the natural and cultural heritage. However, these tendencies are far from universal, with many significant exceptions being cited earlier, in part a function of the context in which the development occurs. Moreover, both the public and private sectors can be very heterogeneous in their composition. A trend towards increasing scale of tourist enterprises has also been identified but again, because of the service nature of the tourist industry and the diversity of its different sectors, a place for smaller entrepreneurs alongside larger operators seems assured.

Given its composite and multi-faceted nature, examination of other aspects of tourist development often initially involves a disaggregate approach, one in which the elements and agents of development in a particular area or with regard to a specific form of tourism are identified and their interrelationships established. Such an approach is utilized in succeeding chapters. In Ch. 3 an attempt is made to establish on this basis general processes and typologies of tourist development. Ch. 4 highlights the diversity of tourist demand while Ch. 5 examines the wide range of resources which might be developed for tourism and techniques for drawing these together in an overall evaluation of tourist potential. Recognition of the composite nature of tourism and the multiplicity of players involved in its development is also essential if a more balanced and holistic approach to impact assessment (Ch. 6) and planning for tourism (Ch. 7) is to be attempted.

3

Processes and typologies of tourist development

To date, relatively few writers have tried to identify and clarify different types or processes of tourist development along the lines of the general models discussed in Ch. 1 (Figs. 1.3, 1.4 and 1.5). Much of the literature on tourism is ideographic in nature, with few attempts being made to compare case studies let alone generalize from them. The typologies that have been proposed serve the useful purpose of highlighting the fact that different processes of development can and do occur. More importantly, the criteria used in deriving these typologies can provide a useful means of analysing tourism in other situations for as Préau (1968, p. 139) points out: 'A rigorous classification is less important than an analytical method of examining reality.' Although the criteria used vary, the typologies generally take into account the characteristics of the developers and the resource being developed, the context and sequence of development and its spatial organization. Particular attention will be paid in this chapter to various features highlighted in Chs. 1 and 2, notably the sequence and context of development and the changing role of local and external developers. The latter is seen to be especially important in terms of some of the broader developmental issues raised in Ch. 1, notably issues of development associated with greater degrees of self-reliance and independence. An attempt is made to match changes in the control structure with changes in the tourist markets served but detailed analysis of the evolution of demand is often lacking in many of the studies reviewed.

Most typologies have been confined to local and regional developments in particular environments. This chapter is accordingly structured in this manner. Reviews of processes and typologies of tourism in coastal areas, ski-field development, rural tourism and urban tourism in developed countries are followed by a systematic examination of tourist development in developing countries. Conclusions are then drawn.

Developing countries here are broadly defined as all those States outside of Europe, North America, Japan and Australasia. The term 'Third World' is also used synonymously with developing countries. Developed/developing countries also generally correspond with the centre–periphery definition of Hoivik and Heiberg (1980) whose study allows the two sets of countries to be put in some global perspective. They estimated that in 1971, four-fifths of all international tourism was between centre countries and one-twentieth between periphery countries. Centre–periphery movements accounted for a further one-twentieth with the remaining tenth going from the periphery to the centre. Domestic tourism in centre countries is also much stronger than in peripheral ones.

The relative importance of the different types of tourism outlined above varies from one developed country to another, depending amongst other factors on their size and physical diversity. Some perspective on their importance for European holidaymakers is provided by a survey in the European Community (European Omnibus Survey, 1986) which showed that 52 per cent of respondents had spent holidays during 1985 at the seaside, 25 per cent in the countryside, 23 per cent in the mountains and 19 per cent in towns (multiple responses were given).

Considerable regional variation also exists in the extent to which researchers have examined the processes of development for each of these types of tourism. European researchers, particularly geographers, have laid greater emphasis on development processes than their American counterparts who have focused more on resultant impacts. When the latter have examined development processes it has often been in the nearby developing countries of the Caribbean. Consequently, much of the developing country literature on tourism is based on small island examples to the neglect, until comparatively recently, of tourism in some of the larger countries of Africa and Asia. Tourism in South America remains largely an untapped area of research while work done in other parts of the world, for example Eastern Europe and Japan has yet to be diffused into the Western literature.

COASTAL TOURISM

Coastal tourism is undoubtedly one of the most significant forms of tourism today with domestic and international tourist flows in many countries being dominated by a massive summer migration towards the sun and the sea (Pearce, 1987a). The coast, at least in Western Europe, is also a well established destination (Fig. 1.5). Amongst the features which characterize tourism there from other areas is the importance of natural resources – sun, sand and sea. Comparatively little investment is often required to exploit these as attractions, with tourist activity being generally informal and unstructured and much of the development being concerned with providing access and accommodation. The nature of the environment, and variations in its physical characteristics, have also influenced the form that development has taken. Much research on coastal tourism has involved morphological studies of resort form and function but increasing attention is being directed to underlying processes.

The seminal typology on tourist development along the coast was Barbaza's regional study of the Mediterranean–Black Sea coastline. Barbaza (1970) distinguished three types of development using the following criteria:

1. The size and extent of the existing population and the vitality and diversity of its activities before the introduction of tourism.
2. The spontaneous or planned nature of the tourist facilities provided.
3. The localized or extensive nature of the tourist area.

Spontaneous development: Costa Brava–Côte d'Azur

In both these cases, tourism developed spontaneously. Along the Côte d'Azur (French Riviera), tourism developed in two stages. The first was characterized by a winter influx of the well-to-do in the eighteenth and nineteenth centuries which gave rise to resorts such as Cannes and Nice and the construction of villas on the slopes backing the littoral. In a second period, after the Second World War, mass summer tourism developed. This was accompanied by a general movement downslope to the beach and by massive and anarchic ribbon development of the littoral between the existing urban centres.

A comparable rocky coastline with a limited hinterland is found on Spain's Costa Brava. The aristocratic phase of tourism was largely absent here, the region being one of small fishing ports, some agriculture and a little industry associated with cork. Such activities created little functional unity and few links were developed throughout the coast or with the inland cities. However, the coastline and climate were extremely attractive and this, coupled with the relatively cheap cost of living, led to an influx of summer tourists in the post-war period. Demand preceded supply, however, the region being ill-equipped for such activities. In the race to develop, much construction was anarchic. Some planning measures were introduced but environmental degradation was not altogether unavoided. There was also a substantial spatial reorganization of the region. Tourist facilities developed the length of the coast, the infrastructure was modernized and links were developed not only with the larger region but also the rest of Spain and, to a certain extent, Europe itself. A certain functional unity emerged. Nevertheless, many of the traditional activities continue.

Tourist resorts resulting from a planned and localized development – the Black Sea Coast

Generally sandy, flat and low-lying, the Black Sea littoral of Romania and Bulgaria was dominated by three large ports (Constanza, Varna and Bourgas) with little population or activity outside these centres. In the post-war period, the socialist governments embarked on a programme of tourist expansion to improve the inflow of foreign exchange and to promote social tourism. The decision to develop tourism was a conscious, carefully calculated one (the market was analysed, the capacity of beaches assessed) preceding virtually any tourist activity. This, coupled with the collective ownership of the land and the State's role in financing, led to the rapid construction of large holiday complexes of 15,000 to 25,000 beds, such as Mamia in Romania and Zlatni Pjasac in Bulgaria. Functional and very localized, such resorts have scarcely had any effect on the previous organization of the region which continues to be dominated by the large ports.

Extensive development: Languedoc–Roussillon

A number of small local resorts developed on France's Langue-doc–Roussillon coastline but prior to the massive development operation of the 1960s the tourist potential of the coastline was largely unexploited (see Ch. 7). Although the object of a development plan, the Languedoc–Roussillon operation differs from the Black Sea projects in that the plan not only incorporates the construction of new functional complexes but also the expansion and redevelopment of existing centres. Moreover, these are linked together by the infrastructure which thus unifies the region.

Significant differences emerge in Barbaza's study. In the first case development is dictated by demand, in the two others by supply. In the Costa Brava example, it is a question of controlling rampant private development; in the other two, the State plays a vital role in developing *ex nihilo* functional resorts. However, the private sector also takes a major part in Languedoc–Roussillon.

Other recent writers, while not proposing typologies as such, have underlined factors resulting in differences in the degree and type of tourist development which have occurred along the Costa Brava and brought about variations on the broad regional theme introduced by Barbaza. Morris and Dickinson (1987) emphasize the importance of the Costa Brava's topography – one of coastal hills, cliffs and small bays – which has prevented uniform strip development by restricting access, limiting building sites and rendering difficult the provision of water and other supplies. Variations from resort to resort and a punctiform pattern of development also result from the high degree of local ownership, with Morris and Dickinson citing a 1982 survey showing only eight of 161 hotel firms along the coast were in foreign ownership. Further differentiation is introduced by the political and administrative situation, with greater fragmentation existing here than along the southern coast of Spain – twenty-two municipalities occupy the Catalan coast. Such fragmentation contributed to 'an unconscious differentiation of the area' as under Franco land use planning was not in the hands of the region but in those of the municipalities who (p. 19):

> guided it in different directions according to dominant local interests
> – those of major landowners, or more frequently those of mayors and
> other municipal officers who were also proprietors and perhaps owned
> construction companies . . . local government held the strings and could
> control also the level and direction of corruption. Some municipalities
> like Begur and Palafrugell, developed on a restructured basis, with villas
> and small hotels for a discriminating clientele. Others welcomed massive
> new construction, as did Lloret and Rosas from the late 1960s.

Priestley (1986) also attributes the rapid growth of Lloret, now the largest resort of the Costa Brava, to the *laissez-faire* attitude of that municipality and compares it to the stricter building codes enforced in Tossa and to the more moderate tourist expansion at Blanes which has had a broader economic base which includes agriculture, textile industries and fishing (Plate 12).

Priestley's detailed aerial photographic analysis of change in the three resorts over the period 1956–81, shows growth in tourism there has taken two main forms: expansion of a hotel and apartment zone around or adjacent to the existing urban nucleus and the development of *urbanizaciones*, sprawling subdivisions of villas (Fig.3.1). Many of the clients for the former come in packages offered by tour operators while the *urbanizaciones* are filled more by independent holidaymakers (Cals, Esteban and Teixidor, 1977). Elsewhere along the Costa Brava new forms have appeared, notably on the low-lying land at Ampuriasbravas which has been developed since 1970 as a marina or *cité lacustre* to cater for a growth in recreational boating.

A similar mix of *urbanizaciones* and larger resorts comprised of hotels and apartments catering predominantly for the package-tour market also occurs further south along the Costa Blanca (Dumas, 1976) and Costa del Sol (Mignon and Heran, 1979). Again, local or regional

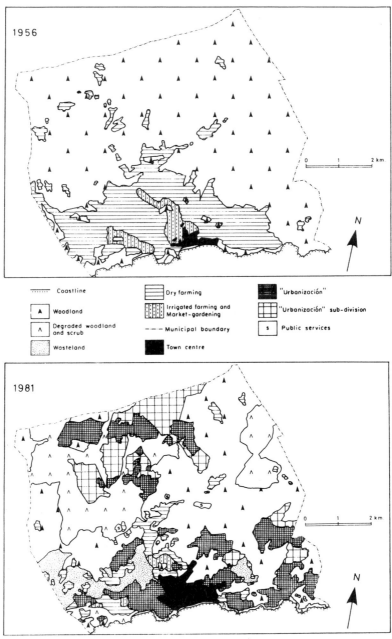

Fig. 3.1 Land-use changes in Lloret de Mar (1956–81)
(Source: Priestley, 1986).

variations are found. Development of the coast south of Alicante to Cabo de Palos is more recent than that part of the Costa Blanca to the north. Development there has taken the form of a series of large Centros de Interes Turistico Nacional (CITN), resort enclaves in which the activities of the tour operators have been linked with those of real estate promoters. As CITNs, these developments benefit from generous credit facilities and fiscal assistance from the State (Vila Fradera, 1966). With the exception of Santa Pola del Este which has been grafted on to a fishing port with some established holiday homes, the others – La Zenia, Dehesa de Campoamor and La Manga del Mar Menor – have been developed *ex nihilo* on large agricultural holdings at the initiative of the land owner. Dumas (1975a, p. 48) describes the development of La Zenia and Campoamor thus:

> The property along the coast (between the road and the sea) is subdivided and several apartment buildings and bungalows and villas are built. Other serviced sections are put up for sale and to rapidly attract clients [and to give some life to the resort] a large hotel is built and filled by tour operators. In the same way, a large number of apartments are rented through German, Dutch and French agencies.

The process at La Manga del Mar Menor has been similar but on a larger scale and with more tour operator involvement (Dumas, 1975a).

A less benign view of foreign tour operator participation in Spanish hotel construction is presented in the pioneering study by Gaviria *et al.* (1974). The process outlined by them is one in which residents of coastal resort areas, often with few resources other than their land, obtained a significant share of hotel construction costs by apparently advantageous loans (at little or no interest) from the tour operators. In return, the hoteliers contracted their rooms for from 5 to 10 years thus securing not only their investment but also their business from the tour operators. However, the contracted room rates were inevitably heavily discounted, with agreed price increases subsequently falling well short of rates of inflation. In this way, the local hoteliers were tied into a system of dependence on the foreign tour operators who benefited from guaranteed accommodation at very little risk and at low rates. Ownership may essentially have been local but control, at least in the boom years of the late 1960s, was very much vested in the hands of the foreign tour operator. Gaviria *et al.* are also very critical of the degree of foreign investment in *urbanizaciones,* particularly in the Canaries, and go so far as to speak of a *produccion neocolonialista del espacio de calidad.* In other words, quality Spanish resources are being exploited and consumed to meet the leisure needs of Western Europe, with much of the profit coming from such development flowing back to the industrialized countries of Europe.

Other European writers have looked at differences in the

location and development of resorts in specific coastal regions without elaborating at length on the underlying processes, particularly the agents of development. Examples here include Casti Moreschi's (1986) study of the Venetian littoral and Clary's (1984) work on the spatial organization of coastal resorts in Lower Normandy. Some interesting contrasts between the development of Languedoc–Roussillon and the Baltic coast of the Federal German Republic have been made by Moller (1983). Research on British coastal resorts appears to have been confined mainly to morphological studies of the more traditional resorts (e.g. Robinson, 1976). An exception to this is Stallibrass' (1980) study of the structure of tourism in Scarborough in which she notes the overwhelming dominance of owner–managers in the hotel sector and the role of local government.

Morphological research has also been undertaken in the United States (Stansfield and Rickert, 1970). Case studies of the development of Atlantic City, New Jersey (Stansfield, 1978) and Grand Isle, Louisiana (Meyer-Arendt, 1985) have also been undertaken in terms of a resort cycle, the former contributing to the elaboration of Fig. 1.4; the latter being interpreted in terms of it. The only American attempt at classifying coastal tourist development appears to be that by Peck and

Table 3.1 Peck and Lepie's typology of tourist development

Rate of change	Power basis	Pay-offs and trade-offs effect on life-style of community
Rapid growth	'Bedroom communities. Summer residents Specialized commerce. (Outside financing.)	Rapid change of local power. New power structure and economy.
Slow growth	Individual developments. Local ownership. Expanding local commerce. (Local financing.)	Slow change of norms. Stable power structure. Expanding local economy.
Transient development	Pass-throughs. Weekenders. Seasonal entrepreneurs. (Local financing.)	Stable norms. Individual mobility within power structure and economy. Little overall change in local economy.

(*Source*: Peck and Lepie, 1977, p. 160).

Lepie (1977) whose typology was developed for a study of small coastal communities in North Carolina. Three major criteria were established:

1. The rate of development, encompassing both magnitude and speed.
2. The power basis, which includes land ownership, source of financing, local input and the relation of local traditions to the development projects.
3. The impact of the host communities as expressed in terms of 'pay-offs' (e.g. benefits to the host culture) and 'trade-offs' (primarily the social impact).

Using these criteria a threefold typology was developed (Table 3.1) and illustrated by three North Carolina communities. Although the differences in the examples do not appear as pronounced as the typology suggests, this does, nevertheless, provide a useful framework for the study of tourist development. The *rapprochement* of the effects of tourism and the processes of tourist development is especially useful.

SKI-FIELD DEVELOPMENT

Ski-field development, like coastal tourism, depends on the exploitation of natural resources. In contrast to the latter, a significant investment is required in the attractions sector to provide the necessary uphill facilities. This, coupled with difficulties of access and sparse populations in many alpine areas, has led to much outside participation, though not necessarily international involvement, in the development process.

The literature on ski-field development is heavily oriented to studies from the European Alps, the world's most developed alpine region in terms of skiing. Barbier (1978) calculated that in 1975–76 about 60 per cent of the world's 20,000 ski-lifts were located in the European Alps. In his global survey, Barbier stresses the different social, physical and economic contexts in which winter sports have developed and how this has given rise to different types of ski resorts. This section reviews several typologies of French ski resorts, then broadens the discussion to a consideration of processes in the European Alps in general.

In a first classification of ski-field development in the French Alps, Préau (1968) concluded that in any situation three sets of factors intervene:

1. The state of the local community when development begins – its size, its dynamism, its facilities.
2. The rhythm of development – whether this coincides or not with the growth possibilities of the local community.
3. The characteristics of the site and the technical and financial possibilities for developing it.

These ideas are developed in a second article (Préau, 1970) where he proposes two different scenarios for the development of tourism in alpine areas (Fig. 3.2). Between these two extremes, Préau recognizes the existence of a number of intermediate situations.

Chamonix – nineteenth century

The first scenario refers to summer tourism in the nineteenth century.

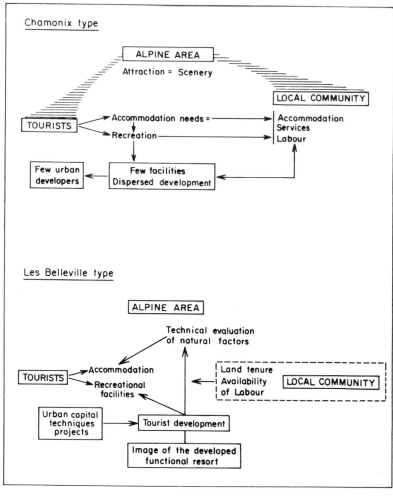

Fig. 3.2 Préau's scenarios of alpine tourist development.
(*Source:* after Préau, 1970).

The diagram reads from the top to bottom and emphasizes local conditions and factors which have been modified by tourists discovering the attractions of the alpine environment. The local society adapts rather readily to a diffuse and easily accommodated tourist demand. Urban developers from outside the area play only a gradual and complementary role, providing, for example, large hotels or capital for a mountain railway. Tourism is an effective germ for further economic and social development and little by little provides the conditions for 'take-off' (Veyret-Verner, 1972).

Les Belleville – 1970

A completely different process operates in this contemporary situation and the diagram reads from bottom to top. Development now begins with the image of a functional resort conceived by urban promoters. It is no longer the mountains as such which are being presented to the tourist but the developed facilities – apartments, ski-lifts and associated recreational equipment. The mountain is reduced to a technical analysis of its characteristics – capacity of the ski-field, construction possibilities, ease of access, etc. Given the scale of the operation, perhaps in the order of 5000–10,000 beds, it is the finance, techniques and know-how of the city which are applied. The only demands made on the local community are for its land and labour.

Integrated and catalytic development

Building on some of the ideas expressed in these studies, this writer (Pearce, 1978a) has proposed a more general twofold classification based on the division of responsibility in the development process:

1. 'Integrated' development implies development by a single promoter or developer to the exclusion of all other participation.
2. 'Catalytic' development, on the contrary, occurs when the initial activities of a major developer generate complementary developments by other companies or individuals.

This basic difference in the division of responsibility influences not only the nature of the development process but also the form of the resulting resort, its location and, to a certain degree, the type of clientele served.

Integrated development

The concept of the 'integrated' ski resort (Cumin, 1970) is now generally accepted in France where La Plagne remains the classic example. Other resorts exhibiting similar characteristics to La Plagne have developed independently elsewhere – for example, the new marinas along the Mediterranean coastline – and it is possible to enlarge on Cumin's ideas and extend these to embrace a more general process of development characterized by the factors described below.

A single promoter.

The entire resort is developed by a single promoter or company. This one company must have at its disposal adequate financial and technical resources. Consequently, such developments are generally limited to large metropolitan financial concerns. Local participation is thus largely excluded from the development process although some locals may find work in the construction and subsequent staffing of the resort and local authorities may be induced to undertake public works programmes such as roading.

Balanced development.

This unity of management more readily permits, though does not necessarily ensure, effective overall planning and balanced development, both technically and financially. With but a single source of direction, the ratio of ski-lifts to apartments can more readily be maintained and the technical problems inherent in marina construction more easily overcome (Giraud, 1971). A common budget allows those aspects of the project which may be initially unprofitable (e.g. the installation of ski-lifts) to be compensated by more lucrative operations (e.g. the sale of real estate).

Rapid development.

This technical and financial co-ordination facilitates very rapid, yet balanced, development which in turn permits a short-term return on the capital invested.

A functional form.

Co-ordination, coherent planning, and physical integration of the resort's facilities may result in a very functional form whereby the holidaymaker, his habitat and his recreational facilities are brought together in a very localized and close-knit resort. Thus condominiums and hotels cluster at the very foot of the ski-slopes and marinas permit the yachtsman to step out of his back door and on to his yacht.

Isolation.

Complete freedom is necessary to develop such resorts, thus they are commonly located away from existing settlements, on comparatively isolated stretches of the coastline or above the line of permanent settlement in the Alps. This complements the need for integration with the natural elements but further removes the resort from the possibility of insertion in the local milieu.

High status.
The first-rate facilities which such a functional resort offers attracts a high class of tourists. Thus the increased costs which are normally associated with the development of isolated sites (even though the price of land itself may be less) are offset by higher tariffs and the overall financial structure of the operation. Indeed, comparative isolation may even enhance the resort's status.

Catalytic development

Elsewhere a single promoter may dominate the development process without, however, monopolizing it. Rather his activities serve as a 'catalyst' by stimulating complementary developments. This process, which might be termed 'catalytic' development, is characterized by the following steps:

1. Initial impetus comes from a single large promoter, often a major outside company, who provides the basic facilities and the conditions for 'take-off'; the primary attractions (ski-lifts, thermal baths, boat harbour), major accommodation units (large hotels, condominiums), publicity and promotion.
2. The success of these activities engenders a spirit of confidence, creates a new demand and encourages the development of complementary facilities: secondary recreational facilities (night clubs, bars, cinemas, bus excursions, mini-golf, etc), alternative accommodation (chalets, small hotels, *pensions*, furnished rooms) and shops. These projects require more modest investments, thus permitting the active participation of smaller local companies and individuals in the development and management of the resort.
3. The expansion of the resort now depends essentially on the operation of a free market system with both the principal promoter and the secondary developers providing facilities to meet the demand. However, if these secondary developers do not respond sufficiently then the principal promoter will have to step up his own activities in order to safeguard the profitability of his existing operations. Conversely, it is also essential to guard against excessive speculation and overdevelopment. In some cases the principal promoter may impose a predetermined programme on the secondary developers, in others planning regulations or the judicious intervention of the local authority may effectively control growth.

The resulting resort differs significantly from the integrated one. Firstly, catalytic developments are usually grafted on to existing settlements. There, however, the major projects often locate some distance away from the centre around which are concentrated the activities of the locals on their existing property. Secondly, the presence of existing dwellings, together with the multiplicity of developers and the less intensive nature of their projects, gives rise to a much more diverse and less concentrated resort than that which results from integrated development. The range of accommodation types offered also broadens the base of the resort which may attract several different classes of visitor.

The characteristics of these two processes will be illustrated in more detail with reference to two ski-resorts, La Grande Plagne and Vars. Similar processes are, however, evident in other environments. Port Grimaud and the Marines de Cogolin, marinas on the gulf of St Tropez (Fig. 7.7) typify integrated developments on the coast while the expansion of Gréoux-les-Bains, a spa in the Alpes de Haute Provence, parallels the catalytic process experienced at Vars (Pearce, 1978a).

La Grande Plagne

The development of the Bellecote–Montjovet massif in Savoy, France, into the tourist complex known today as La Grande Plagne began in the early 1960s. Faced with declining agricultural returns and the closure of local lead mines, five communes in the Isere valley joined together in a '*syndicat intercommunal*' to promote the development of their mountain resources. Although the communes built an access road to the chosen high altitude site (at 2000 m), the first developer, a modest regional company, soon failed financially. Subsequently a group of banks, mainly based in Paris, established a development company, the SAP (Société d'Aménagement de La Plagne), and entered into a formal contract with the communes in December 1961. Basically this contract specified that the SAP would have the exclusive rights to construct a network of lifts and accompanying accommodation at a predetermined rate while the communes would concede the land occupied by the ski-field for 30 years for a small percentage of the lift revenue and would make available for sale the land on which the resort would be built.

Construction of La Plagne, the base resort, began the following year. Several levels of integration can be seen in the development of La Plagne, widely acknowledged as the first 'integrated' resort. Firstly, La Plagne was developed entirely by the SAP. By 1968 the company had built 5000 beds, mainly in condominiums, with a lift network having an hourly capacity of 9000 skiers. While the operation of the lifts was initially unprofitable, the sale of apartments, block by block, assured a rapid return on capital invested, permitting the self-financing of the whole operation. The SAP also managed or subsidized certain shops and hotels as well as the cinema during these first years as their presence

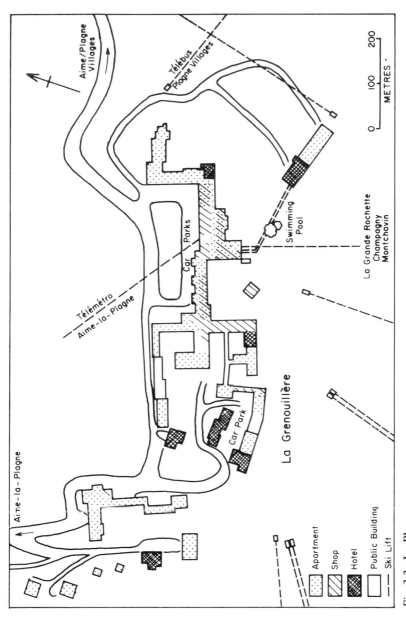

Fig. 3.3 La Plagne

was essential to attract would-be buyers. However, it is in the form of the resort that integration is the most evident. All the accommodation is located at 2000 m, high above the traditional settlements situated in the valley below, on a small shelf towards which converge the principal *pistes* and from which leave the major lifts (Fig. 3.3). Moreover, the amount of accommodation has been calculated as a function of the capacity of the ski-field to ensure that the benefits of this functional association are not marred by overcrowding of the slopes. There is also a virtual physical integration of the thirty or so buildings that constitute La Plagne. Use of motor vehicles within La Plagne is consequently much reduced while those vehicles arriving at the north side of the resort are effectively separated by its linear form from the skiers on the south side. This integration also limits the amount of land to be purchased, reduces servicing costs, enables a more rational use of central heating and permits easier maintenance and management.

Plate 2   Integrated development: at Aime La Plagne 'integration is carried to the extreme as a single, gigantic monolith accommodates not only 2500 people but also houses a cinema, discotheque, restaurant and a full range of shops'

With La Plagne established, the SAP progressively developed four smaller satellite resorts nearby at a comparable altitude. In one of these, Aime La Plagne (or Aime 2000), integration is carried to the extreme as a single gigantic monolith accommodates not only 2500 people but also houses a cinema, discotheque, restaurant and a full range of shops. However, there is a move away from this total integration in the subsequent satellites such as Plagne Villages. This reflects changing market demands although the clientele for these satellites continues to be drawn from the same professional classes as for La Plagne itself, particularly from the *Polytechniciens* from Paris.

Other than employment in some of the new jobs created, the local population has been effectively excluded from the development

Plate 3   Catalytic development: the gradual expansion of ski-field capacity has resulted in the construction of chalets, small hotels and a new school in the hamlet of Vars Ste Marie

of La Plagne and the other high-altitude satellite resorts. However, the opening up of the immense ski-field by the SAP has enabled the communes to undertake more traditional developments (small hotels, chalets, holiday camps, *gîtes*) around existing settlements further down the mountain (Longefoy, 1550 m; Montchavin, 1300 m; Champagny 1250 m). These are joined to the upper ski-basins by a series of linking lifts and are being developed as part of the overall complex of La Grande Plagne.

In this broader context the high-altitude integrated resorts may be thought of as having had a catalytic effect on the area as a whole. The question of scale is therefore very important in assessing tourist development. By 1974 La Grande Plagne had an accommodation capacity of 14,000 beds, forty-five lifts and eighty ski-trails totalling 135 km. An estimated 500,000 skiers visited the complex during the 1973–74 season.

Vars

The catalytic effect of the intervention of a large outside company is clearly evident at Vars (Hautes Alpes). Although the commune's potential as a major ski-resort was recognized before the Second World War, the few hundred Varsins, for the most part small farmers, lacked the means to develop more than four or five simple lifts and a few small hotels and chalets in the immediate post-war period. With local agriculture in decline, the commune was stagnating and suffering steady population losses through out-migration. This situation changed in 1958, with the election of a new mayor, a former politician from Paris. The new mayor was able to interest a group of Paris-based financial concerns in forming a company, the (Société pour l'Equipement et le Développement de Vars) SEDEV, to develop Vars as a tourist resort. Unlike the local population, the SEDEV had both the financial and technical resources to create a network of lifts large enough to launch the resort on a national scale. As the SEDEV constructs each stage of the network the commune, in the terms of a formal contract, cedes a specified area of land to the company for the development of accommodation. The SEDEV has limited itself to large-scale condominium development, leaving open the possibility of alternative forms of accommodation. Previously reluctant to take the initiative, the Varsins responded enthusiastically to the signing of the agreement with the SEDEV in 1962. From this date the register of building permits shows a marked increase in local activity; new chalets are built, small hotels and *pensions* are opened or enlarged, a shop-window is improved here, a new shop built there, snack bars and *boîtes de nuit* appear. Most of this activity, assisted by loans from government agencies such as the Crédit Hotelier and the Crédit Agricole, has been concentrated around the existing hamlets (Ste Catherine, Ste Marie and St Marcellin) while the SEDEV and other non-local companies have developed a new centre at Les Claux (69 per cent of tourist beds in 1972), some 3 km up the valley. The municipality was responsible for purchasing the necessary land at Les Claux and for servicing and subdividing it. Other new secondary roads were also built, together with an improved sewerage system. The municipality also entered directly into the tourist industry with the construction of accommodation in the form of *gîtes*, hostels and apartments. Throughout, it has had to rely on extensive loans from central government.

By 1972, 10 years after the contract with the SEDEV had been signed, Vars had an accommodation capacity of more than 8600 beds and a network of twenty-five lifts. Much of the resort's success has clearly been due to the strong and able character of the mayor who has been able to balance local interests with those of the external developers and control the rate of growth through the judicious granting of building permits and the application of the municipal building code.

The ski-resorts of the French Alps, or at least the large integrated ones, are placed in the broader context of the alpine arc in a very useful and comprehensive article by Barker (1982). Barker identifies differences in the scale, intensity and form of tourist development between the western Alps (France, western Switzerland) and the eastern Alps (eastern Switzerland, northern Italy, Austria, Bavaria). Figure 3.4(a) shows that in the western model large integrated ski-resorts have been built in the subalpine zone long after the local population had retreated to the main valley where agriculture, forestry and manufacturing provided employment. The forested zone acts as a buffer between the valley and uplands, separating the valley communities from the high-altitude ski-resorts. In contrast, tourism in the eastern Alps (Fig. 3.4b) coexists with a strong pastoral economy (p. 408):

> hotels, pensions, and vacation houses arose around the traditional valley communities, many of which have ski and summer-sport facilities. . . . Above the protective and protected forests overlooking the valleys, the old pasture huts were converted to weekend retreats by villagers who held grazing rights to these lands. With the spread of development, additional ski runs were cut into the forested slopes . . . and rack railways and gondolas enabled summer tourists to reach the mountain peaks. . . . Where communities invested in a summer skiing project roads lead through the alpine zone to glaciers on which ski lifts were installed.

In the French case, where the Alps had experienced a prolonged period of rural depopulation, the main thrust for development in many areas came from more distant urban capital, as was noted earlier. Moreover, the development of high-altitude integrated resorts had been vigorously supported by the State through Le Plan Neige

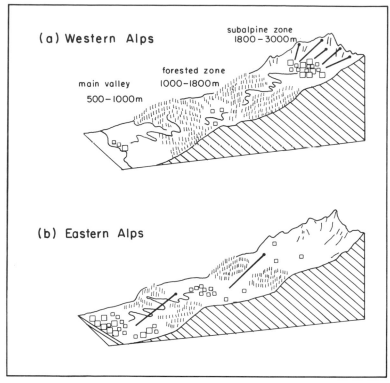

Fig. 3.4 Barker's model of tourist development in the European Alps. (*Source:* redrawn from Barker, 1982).

(Barbier, 1978; Knafou, 1979). Elsewhere in the Alps much of the impetus has come from within strong rural communities with a tradition of local autonomy in planning which has favoured community-based investment initiatives. Barker also points (p. 408) to 'the preference of German-speaking tourists for participation in local culture rather than vacationing in enclaves', and notes that by the late 1970s more than one-third of the total tourist bednights in the Tirol and Alto Adige were spent in private quarters.

Other recent writers have widened the debate on ski-field development to include several of the dimensions of development raised in Ch. 1 (Dorfmann, 1983; Guérin, 1984). Guérin identifies two main development themes regarding tourism in the French Alpes du Nord: development (*aménagement*) as the exploitation of resources (*mise en valeur*) and development as conservation of the Alps (*sauvegarde de la montagne*). In the first, the emphasis is on investing in and exploiting the natural resources of the Alps to satisfy a growing external demand and to make a profit out of doing so. The implantation of the new resorts which such development involves brings about a major restructuring of the local economy and society. The conservation theme, on the other hand, stresses retaining the existing population, its traditions and values and giving the local residents a fair chance at determining their own future. While these two themes are not mutually exclusive, a certain tension exists between them. Differences in the two are manifested geographically in the ways in which the policies have been pursued. The *mise en valeur* doctrine has favoured the higher altitudes, with the emphasis on the commune as the unit of development. The *sauvegarde* policy, on the other hand, has been oriented more to the *moyenne montagne* and to organization and planning at the level of the micro-region (*pays*). Guérin suggests that while the former policy has received the strongest support during the 1960s and 1970s, some swing to the *sauvegarde* strategy has occurred more recently with the Socialist victory in 1981 and the economic recession of the early 1980s.

RURAL TOURISM

Various factors attract tourists to rural regions; natural features such as rivers, lakes and wooded areas, or cultural/economic ones, for example picturesque villages or the cultivated countryside. Tourists spend their holidays in the countryside as a change from urban areas, seeking the peace and relaxation which a rural environment can offer. Others come for more active pursuits: fishing, walking, boating, etc. Rural vacations tend to be informal and unstructured; travel agents and tour operators play little if any role in their organization and the dominant mode of transport is the private car. Tourist development in rural regions is characterized by a multiplicity of small-scale, independent developers and diffuse developments. The local authority is usually the most important unit of government with its policies towards tourism and its power to encourage or limit construction being a critical element in the development process. Research on rural tourism has focused on second homes and farm tourism, with other forms of tourism, for example those involving camping and holiday villages being dealt with less frequently (Mormont, 1987).

Second homes

The literature on second homes is very extensive but much of it has been descriptive, examining in particular the distribution of second homes and their impact (Clout, 1972; Coppock, 1977). Nevertheless, several writers have concentrated on the process by which second-home settlements emerge and have proposed general models of such processes.

Lundgren (1974) has produced a three-stage model based on the Canadian experience, showing changes in the spatial relationships between the urban centre and a second home or cottage region as the urban area expands (Fig. 3.5). In the first phase, demand from a medium-sized centre has generated a small second-home region, typically in an area of broken relief or around a body of water. As the urban area grows, so the demand for second homes increases and the second-home region expands, mainly away from the city (Phase II). Lundgren suggests that 'the inside expansion is more urban in character, whereas the outside push still retains the features of the typical vacation home development'. In the third and final stage of the sequence, the original second-home region becomes engulfed by the expanding metropolis and now forms a part of the city itself, with the former second homes being transformed into permanent residences. Meanwhile a new, more distinct second-home region has developed, for the demand for weekend or vacation accommodation has not abated but rather increased. This outwards expansion and growth in demand is not only a function of the larger population, but development of highways, increased car ownership, greater leisure time, and a desire by many local authorities to increase their tax take through intensification of land use as well as the activities of real estate developers promoting speculative subdivisions.

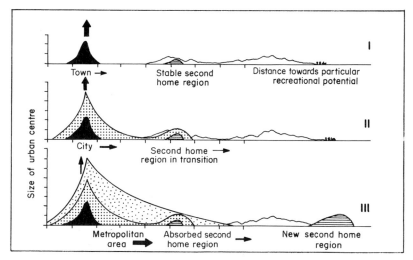

Fig. 3.5 Lundgren's model of urban expansion and second home regions. (*Source:* redrawn from Lundgren, 1974).

Plate 4 Purpose-built second homes, Alpes de Haute Provence, France

A similar three-stage process has also been identified by Boyer (1980) in the Paris basin. In the first stage, existing rural dwellings are acquired by city residents, mainly for use as second homes. This is followed by a period of new construction on land bought from local inhabitants. At the beginning of this phase, second homes coexist with primary residences but the latter soon become dominant. In a third phase, individual activity by both city and local residents is replaced by larger subdivisions created either by the local authorities, seeking to increase their tax base, or property developers after a profit. Primary residences now become the norm as the peace and isolation sought by the second-home owners has disappeared. The latter group move further outwards and the process is repeated.

Other aspects of this process have been noted by Lefevre and Renard (1980) in the west of France, in the Marais de Monts region. There a change was observed in the social and geographic origins of the second-home owners during the second phase. Whereas many of the former farm dwellings became the second homes of the family of rural emigrants or nearby urban residents, many of the purpose-built second homes belong to Parisians or residents of the Loire valley. For rural France as a whole, Cribier (1973) estimated that in 1970, half of all second homes had been inherited, a further 40 per cent were former dwellings which had been purchased and one in ten had been purpose built. Lefevre and Renard also point out that by the third phase the local authorities, perhaps supported by some pressure groups, may react to increasing urbanization not by encouraging further expansion but by introducing land-use zoning and other measures to protect agricultural land and limit or direct urban growth.

From the detailed studies by Bromberger and Ravis-Giordani (1977) of a variety of rural communes in the Provence-Côte d'Azur region in the south of France, it is evident that many local variations on this general process are possible with the role of the local authorities being critical. Bromberger and Ravis-Giordani suggest Banon is typical of the early stages of development (Fig. 3.6). The emphasis is on second homes, whose construction or maintenance generates work for local builders. The social–political effects of the second-home owners is minimal although some tensions start to appear between different sectors of the community. As the conflict of interest between local residents wanting to remain in agriculture, other locals wanting to take advantage of new tourism demands and second-home owners (who have local voting rights and may 'take over' the council if they become numerous) have emerged in other communes where the process is more advanced, different policies have been put forward. For example, in Savines, on the shores of Lake Serre Ponçon created by hydro-developments, the Socialist mayor has encouraged the development of a social tourism centre rather than the proliferation of second homes. Such a policy was

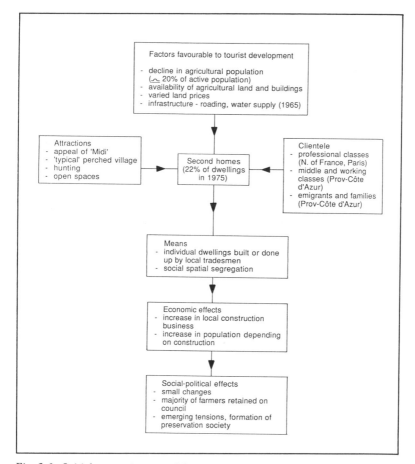

Fig. 3.6 Initial stages in second-home development, the case of Banon, France.
(*Source:* after Bromberger and Ravis-Giordani, 1977).

designed to obtain assistance with developing the shores of the lake for recreational purposes, to create more permanent jobs and to bring into the process a major developer who was not profit-driven so that control could remain in the hands of the local community. At Le Beausset, on the other hand, where agricultural decline had been more severe and where a key group of local property developers influenced municipal policies, second-home expansion was actively encouraged through land-use zoning. At Saint-Cyr, a commune on the Mediterranean coast, tourism has been established for a much longer period and already a marked differentiation occurs between the tourism-oriented littoral, old and new urban centres and the inland vineyards. Pressures and prices on the littoral have been such that tourist development no longer involves the conversion of agricultural lands for tourist purposes but the replacement of one form of tourism by another. Villas, hotels and both small and large properties are being bought up and redeveloped for vacation villages and tourist apartments. The cost of these operations is such that they are limited to large property development companies.

Rates of second-home ownership in the United Kingdom (2 per cent of households) are much lower than elsewhere in Western Europe (France, 16 per cent; Sweden, 22 per cent) (Shucksmith, 1983) and the literature on their development is less extensive. British writers have focused on some of the conflicts which may emerge, particularly the competition for housing between local residents and second-home purchasers, and on the role of local authorities in reducing these (Shucksmith, 1983; Bouquet and Winter, 1987). Some authorities have attempted to introduce 'local needs policies' on new construction but these have often been counter-productive, forcing up prices on existing housing. In the absence of planning legislation, access to housing stock may be controlled by key individuals within the community (Ireland, 1987).

Space constraints and rural policies appear to have prevented the widespread growth of rural second home subdivisions in the United Kingdom comparable to those found in France and Spain. Such commercialized subdivisions were common in the United States in the 1960s and early 1970s when demand for that type of facility declined dramatically, a result of 'shady' business practices and market resistance associated with the economic recession, high energy costs and the emergence of new real estate possibilities such as those provided by time-sharing (Boschken, 1975; Stroud, 1983).

Farm tourism

Whereas second-home development generally involves a transfer of land and buildings from rural to non-resident ownership, farm tourism represents continuing ownership and active participation by the farmer in, typically, small-scale tourism ventures. In Europe, the term 'farm tourism' 'applies only to operations, where, besides the hospitality function, active agriculture is also providing income to the host family' (Dernoi, 1983, pp. 155–6). 'Vacation farm' is the term used in the United States to refer to 'an active, working farm on which extra rooms in the home or extra houses on the farm are rented to guests' (Pizam and Pokela, 1980, p. 203).

A wide-ranging review of farm tourism in Europe was undertaken by a joint Food and Agriculture Organization (FAO)/Economic Commission for Europe (ECE) Working Party in 1982 (FAO/ECE, 1982; summarized by Dernoi, 1983). Aspects of European farm tourism have also been examined by Frater (1983). While not focusing specifically on the processes by which farm tourism has developed, these studies provide a good overview of the basic dimensions of the phenomenon.

In Europe, farm tourism takes two main forms, the provision of accommodation in farm premises, or the supply of accommodation on the farmer's land in the form of cottages and camping grounds. The most widespread is the letting of accommodation (commonly bed-and-breakfast), with the second form being more characteristic of northern Europe. Austria is generally recognized as being the country where farm tourism is the most extensive with over 100,000 guest rooms on almost 30,000 farms being made available in 1970. By 1980, one out of every ten Austrian farmers was letting rooms to tourists with rates in such regions as the Tyrol being much higher (28 per cent). In the United Kingdom, at least 10,000 farms were estimated to offer farmhouse accommodation, with a similar number providing self-catering accommodation.

Frater's survey of Herefordshire farmers showed four main factors influenced the farmers' decision to provide accommodation: to increase annual income (35 per cent), to offset falling income from agriculture (20 per cent), to utilize disused resources (16 per cent), and to enjoy the visitors' company (25 per cent). While farm tourism depends essentially on individual initiatives, farmers in many countries are supported by State and other organizations, particularly by way of development grants and marketing assistance. For example, some technical advice for British farmers on establishment costs and considerations for various types of accommodation and resource-based activities is provided by the Dartington Amenity Research Trust (1974) report on farm recreation and tourism. Co-operative marketing is particularly necessary as individual farmers do not have the means themselves to promote their facilities effectively in the urban markets. France, for example, has a well-established 'National Federation of Gîtes', which not only provides advice and assistance to the farmer but operates a national register of approved properties and a centralized booking system.

Co-ordination is especially important in New Zealand, where farm stays are offered not only to independent holidaymakers but also included on some of the more innovative coach tour packages. In such cases, regional co-ordinators of farm-stay organizations are needed to ensure sufficient farmers are available in a given area to cater for entire coach-loads of visitors at one time.

In the United States, farm tourism takes the form of dude ranches, concentrated particularly in the west and vacation farms, found mainly in the east (Vogeler, 1977). Vogeler estimated there were some 2000 such farms and ranches in the early 1970s. Early promotion of the more exclusive dude ranches was often undertaken in conjunction with railways. States with large concentrations of vacation farms have county and State associations which produce listings for tourists. Pizam and Pokela (1980, p. 216) concluded from their study that:

> an integral part of the planning and development of the farm vacation business will be to assist existing vacation farms in developing their marketing and promotional efforts, so they will not be harmed by an influx of new vacation farms with greater promotional efforts, particularly as the supply of vacation farms begins to meet the demand.

TOURISM IN URBAN AREAS

Although many capital cities, metropolitan centres and historic towns and cities are important tourist destinations, urban areas have been relatively neglected by tourism researchers (Pearce, 1987a, pp. 178–97). The work that has been done has been primarily morphological, focusing in particular on the distribution of different forms of accommodation (Burtenshaw, Bateman and Ashworth, 1981; Wall, Dudycha, Hutchinson, 1985; Ashworth and de Haan, 1986). Other writers have examined aspects of planning (Chenery, 1979), demand (Wall and Sinnott, 1980; Jansen-Verbeke, 1986; Pearce, 1987d) and the economic impact of tourism (Liu, 1983). While some of the morphological studies have taken account of changing patterns of accommodation over time, little explicit attention has been paid to processes of tourist development in urban areas along the lines discussed in the preceding sections and no attempt appears to have been made to derive typologies of tourism in towns and cities. Nevertheless, it is possible to identify and describe some of the general features of tourist development in such areas.

Towns and cities, perhaps more than the other areas discussed so far, are multi-functional in nature. This factor conditions the form tourism takes and the way in which it develops. Tourists visit urban areas for a variety of reasons, for entertainment and night-life, to appreciate historical and cultural attractions, to attend major sporting events, to shop or just to enjoy the charm and character of a particular city. In many of these cases, tourists share these attractions with city residents; in others, the attractions, for example historical buildings, are derived from former functions. Many cities also receive a considerable volume of traffic generated by other contemporary functions. Administration, commerce and industry attract large numbers of business travellers while a sizeable resident population will generate a significant number of visits to friends and relations. Conferences and special events will draw other visitors. These functions and attractions will vary from city to city, with many North American cities perhaps being less dependent than European ones on historical and cultural attractions while drawing more of their visitors for business, conventions, entertainment and from visits to purpose-built attractions such as theme parks.

Because of this mix of different functions, tourists in urban areas tend to be less homogeneous in their demands than those in coastal or ski-resorts. One consequence of this is that a wide range of agents of development are drawn, directly or indirectly, into providing services and facilities for the visitor. This is particularly the case with the attractions sector where many of the features which appeal to tourists, for example Gothic cathedrals or art galleries, are not oriented explicitly nor primarily towards them. The new tourist demand may or may not sit comfortably alongside existing functions. In such circumstances co-ordination becomes not only more important if tourism is to develop successfully, but also more difficult given the range of interests and differing commitments to serving the visitor. Profit-motivated commercial individuals and companies involved in the accommodation sector or sightseeing businesses, for example, are complemented by a range of municipal and State agencies, civic trusts, and other non-profit-making organizations responsible for managing cultural, historic and other attractions (Jansen-Verbeke, 1988).

As tourist demand is heavily dependent on other urban functions, so changes in these functions will influence the path of tourist development. As the central location of many city hotels reflects demand generated by other central functions, it follows that changes in the distribution of hotels will result from an expansion of the CBD or the appearance of new centres as well as from a shortage of central sites (Pearce, 1987a). This is particularly evident in cities where the tourist industry has expanded rapidly at the same time that major morphological changes have occurred. Gutiérrez Ronco (1977, 1980) has traced in detail the evolution of hotels in Madrid and shows how there has been a progressive shift from the Puerta del Sol to the north and north-east as administrative and commercial activities have developed in that direction, particularly in the last two decades. A similar pattern occurs in Lisbon, where the Avenida da Libertade links the older, smaller hotels of the centre with the more recent, larger and better hotels in the streets surrounding the Parque Eduardo VII, also an area of recent commercial and administrative expansion. Ashworth and de Haan (1986) have proposed a 'tourist–historic city' model which incorporates the notion of a partial migration of the CBD and the emergence of a 'tourist city' in a zone of overlap between the historic core and the contemporary CBD. They illustrate their model with reference to Norwich; it also would appear to fit well in the case of Madrid and Lisbon.

In some instances, tourist development, especially hotel construction, follows the general evolution of the city. Elsewhere tourism is adopted as a rationale for the preservation of historic quarters or is identified as a key element in urban renewal projects. Ford (1979) cites a range of United States cities where the tourist element has been an integral part of the preservation of downtown cores or the adjacent zone of discard: Charleston, the French Quarter in New Orleans, Pioneer Square in Seattle. He also notes (p. 223) that such a process is not without its opponents: 'Critics of preservation are quick to point out that the fancy restaurants, boutiques, architects' offices and night spots serve primarily a tourist and suburban population. The poor and the helpless, they say, are displaced when an area becomes fashionable and rents skyrocket.' Law (1985) has examined other American examples of the incorporation of tourism in urban revitalization projects, notably Baltimore, Boston and San Francisco. He concluded (p. 527): 'Public initiatives and investment are important if tourism possibilities are to be realised, a point accepted in this most free enterprise of countries. Planning and urban renewal agencies can co-ordinate the resources required to create facilities which will attract visitors. However, the success of the tourist industry will depend on the extent to which private investment is stimulated.'

The English Tourist Board (ETB, 1981, p. 24), in its review of tourism and inner city schemes in England, e.g. St Katherine's Dock in London and Piece Hall in Halifax, noted: 'The primary motive in these cases was conservation rather than the promotion of tourism, tourism-related projects being chosen because they are adaptable and well suited to providing uses for old buildings'. The ETB report (p. 25) also emphasizes the role of the local authorities in such cases:

> The local authority may need to initiate development in areas where the private sector is reluctant to take a leading role. In other areas, they can often influence the development process through their role as land owner and by providing services and infrastructure. In addition, their role as planning authorities gives them responsibility for setting out a planned framework to guide and control development.

At the same time the report advises: 'The involvement of the Tourist Boards, tourism operators and local interest groups will also be essential in formulating realistic proposals.'

Tourism, however, has not been limited to preservation projects. In Paris, for example, planning authorities have sought some diversity in their renewal or redevelopment projects by encouraging hotel construction amongst large office complexes through offering land for hotels at half the rate for office buildings (Cadart, 1975; Pearce, 1987a). The Sheraton (995 rooms), for example, forms part of the Ilot Vandamme renewal project while the Japanese-owned Nikko (778 rooms) is located in the Fronts de Seine redevelopment project. The decentralization of large luxury hotels belonging to international or French chains (Meridien, Sofitel) into the fourteenth, fifteenth and seventeenth *arrondissements* began with the construction of the Hilton in 1966 and has continued since then, leaving many of the older, smaller, family-run hotels in the inner city or around the railway stations. Changes in the importance of various modes of transport have also seen the appearance of hotels located along autoroutes on the outskirts of the city, particularly those belonging to the French Novotel chain, and the growth of 3- and 4-star hotels around both Orly and Charles de Gaulle airports. Similar patterns are also evident in other European cities (Burtenshaw, Bateman and Ashworth, 1981).

What is required now is a more comprehensive approach to research on tourist development in urban areas, one which links the various themes outlined at the beginning of this section with other, newer ones. Morphological studies must now be related more explicitly to demand and to the planning process and the role of the various agents of development explored more thoroughly. Such an approach should not be limited to the accommodation sector alone, it must encompass a longer time dimension than is found in most urban tourism studies and, if typologies are to be formulated, it must also embrace a range of examples from various countries.

TOURISM IN DEVELOPING COUNTRIES

Considerable attention has been focused on tourism in developing countries, particularly in the impact literature (e.g. Peppelenbosch and Tempelman, 1973; de Kadt, 1979; O'Grady, 1981). Tourism in such countries is generally portrayed as being distinctive, yet there have been few attempts to highlight systematically differences between developed and developing countries. This is especially true in terms of patterns and processes of tourist development. Moreover, little attention has been paid to exploring differences between and within the wide range of developing countries which are to be found throughout the world. Much of the literature on tourism in developing countries concentrates on coastal tourism. Less frequently studied is wanderlust tourism centred on natural features, historic sites or exotic cultures (e.g. trekking in the Himalayas, the Mayan culture circuit of Central America). Elsewhere, Third World tourism may be urban-based, as in the case of casino development in Lesotho.

Coastal tourism

Many of the coastal studies in developing countries have focused on the small island States of the Caribbean and South Pacific. Lundberg (1974, pp. 197–8) presents a generalized sequence of new hotel development in developing countries based on the Caribbean experience. The perspective is that of the hotel operator, but the sequence also incorporates aspects of development found in Miossec's model (Fig. 1.3) notably the changing attitudes of decision-makers and the host population:

Phase I

Tourism is seen as being very beneficial economically and the government offers significant incentives to hotel operators (tax holidays, waiving of import duties, etc.). Land speculation has not yet set in and labour is cheap.

Phase II

A halcyon period from 5 to 10 years in which the hotel operator does very well financially due to reasonable construction and operation costs, the availability of local produce at reasonable rates, tax breaks and the ability to repatriate profits.

Phase III

Local resentment grows as the differences in wealth between residents and tourists become apparent. The government begins to take a harder line on tourism as some of the economic costs associated with its development start to emerge (foreign exchange leakages, intersectoral competition). Land speculators start getting rich.

Phase IV

The hotel operator is faced with high or rising costs: the tax holiday period has ended, local produce and seafood prices have been driven up, unionized labour becomes more expensive and labour efficiency may drop off. Tourists may receive poor service or even be insulted. Consequently the number of repeat visitors declines. The government reacts by increasing its tourist promotion budget. Foreign personnel may no longer be given entry permits to work in the hotel industry; attempts at training local staff are made but may not be very successful. Tourism has reached a plateau and an economic recession in the tourist market may cause a sharp drop in tourism to the country.

Phase V

The hotel operator experiences hard times as too many hotels were built during the earlier more prosperous phases. Newer hotels had often been built on more expensive land and taken the form of high-rise buildings built at great cost alongside other high-rises. The investors in the new hotels see they cannot operate them themselves and attach their operations to international chains with their more extensive marketing facilities.

Phase VI

This phase completes the development pattern in developing countries. After making a fine profit during the first years' operations, the hotel operator has to face a number of political and other problems which have emerged as the area has matured. There may even be riots as happened in Bermuda, Curacao and Jamaica. Adequate government planning may have avoided many of these problems.

Other than mentioning the three countries listed in this last phase, Lundberg does not illustrate this model with reference to specific places or events though evidence to support many aspects of it can be found elsewhere, particularly in Bastin's (1984) account of developments in Jamaica. He suggests that the growth of mass tourism on the island dates from the Hotels Incentive Bill of 1968 which offered generous tax concessions for large hotel construction. Major infrastructural improvements and the opening of a hotel school also testify to the Jamaican government's favourable stance towards tourism during the late 1960s and early 1970s. Tourist arrivals continued to expand during this period but started to decline after 1975 as a result of three main factors: increased competition from Europe following the introduction of jumbo jets, adverse publicity overseas during the period of 'democratic socialism' of the People's National Party (1972–80) and hostile attitudes to tourists resulting from local perceptions of servitude and an unequal distribution of the economic benefits of tourism. As a result of the downturn in demand after a period of hotel expansion, occupancy levels fell rapidly and many hotels experienced financial difficulties. Many of the international chains running the large hotels reacted to decreased viability by terminating their leases. The government then stepped in to protect a major source of employment and foreign-exchange earnings. By 1977 the Jamaican government holding company, National Hotels and Properties, was responsible for almost half the hotel capacity on the island and for taking over much of their debt servicing. The new administration which came to power in 1980 has tried to divest itself of all hotels but has yet to be fully successful.

Variations on this process in Tobago are described by Haider (1985, p. 64):

> International standard hotels were built by foreign investors during the 1960s and early 1970s, and only recently have three of the five establishments been bought by nationals. Two factors contributed to this change of ownership: the availability of domestic capital after 1973 [as a result of increased oil revenues in Trinidad] and the expiration of tax

Table 3.2 Summary of hotel characteristics in Tobago

Category	Number of rooms	Origin of owner	Management form and origin	% Domestic clientele	% Package visitors	Winter Prices (US$ day)	Services provided
I International Standard hotels	>30	Foreign or Trinidad	Salaried foreign or Trinidad	<40	<40	Max. $131 Min. $111 Ave. $115	Private bath Air conditioning Pool Restaurants Bar/s Shop/s
II Economy class	14–15	Trinidad or Tobago	Salaried (Trinidad) Family (Tobago)	40–90	<25	Max. $75 Min. $38 Ave. $57	Private Bath Air Conditioning (optional)
III Small local guesthouses	<14	Tobago	Family Tobago	>90	0	Max. $27 Min. $17 Ave. $24	Almost no facilities

(*Source:* Haider, 1985).

holdings and other incentives that had been granted to hotel investors for the first decade of operation. In spite of increased local ownership, top management still comes from metropolitan countries.

Haider also shows significant changes in ownership occur with the nature and scale of development, with native Tobagonians only owning or managing either the smaller economy class hotels or the guest-houses which primarily serve the domestic market (Table 3.2). Rodenburg (1980) depicts a similar pattern of ownership in Bali where the large industrial hotels (hotels of an 'international standard' and having more than 100 rooms) are in corporate Indonesian or foreign ownership while local village ownership is greatest in the small 'craft' enterprises (Table 3.3). Wong (1986) underlines the initial importance of domestic tourism in tourist development along parts of the east coast of peninsular Malaysia and suggests the following sequence of development:

- local popularity of the beach brings the developer's attention to an area suitable for resort development,
- the initial resort is often one that meets the demands of domestic tourists,
- this is then followed by better quality resorts catering to the international tourist.

Most of the resorts on the east coast are relatively recent, dating from the mid-1970s when the region was actively promoted by the Tourist

Table 3.3 Local and foreign entrepreneurship in Bali

| Scale of enterprise | Number | Origin of owner | | | Nature of ownership |
		Local village (%)	Greater Bali (%)	Indonesian/ foreign (%)	
Large industrial	5	0	0	100	Corporate
Small industrial	111	33.3	16.7	50	Corporate/ individual
Craft					
'homestay'	195*	66.7	28.5	4.8	Individual
restaurant	NA†	66.7	33.3	0	Individual
souvenirshop	NA†	66.7	33.3	0	Individual

(*Source*: Rodenburg, 1980).
*Gross underestimate, this figure includes 'homestays' from only four areas.
†The numbers of total craft level restaurants and souvenir shops is not available.

Table 3.4 Classification of beach resorts, the east coast of Peninsular Malaysia

Type	Major features	Examples
1. Planned resort complex	Large-scale development. Important role in regional development. State organization in overall planning. Joint development between State and private companies operating at national and international levels.	Desaru resort complex
2. Individual resort of international standard	Medium-scale development. 'Exclusive' character. Joint development between state and private companies operating at national and international levels.	Tanjong Jara Beach Hotel, Club Méditerranée Cherating, Hyatt Kuantan
3. Individidual resort of national/ local standard	Medium-scale development. Developed by private companies operating at State and national levels.	Pantai Chinta Berahi Beach Resort, Chendor Motel, Simjifah Motel
4. Basic accommodation unit (dormitories, *kampong* accommodation)	Small-scale development. Private local enterprise. Adaptation of local resources to tourist demand.	Pulau Tioman, Kampong Cherating Lama

(*Source*: Wong, 1986).

Development Corporation. Wong proposes a tentative hierarchical typology of resorts there based on the scale and quality of the resorts, the extent of public and private participation and development processes (Table 3.4). Wong's typology would appear to provide a very useful basis for examining and classifying coastal resorts in other developing countries.

A broader, structural perspective on tourist development in the Caribbean has been proposed by Hills and Lundgren (1977), who stress the core–periphery interrelationships between the major North American markets and the Caribbean destinations. Their model (Fig. 3.7) emphasizes the dependence of the latter on the metropolitan centres, not only as sources of tourists but also suppliers of the other resources involved in tourist development (capital, know-how, supplies, etc.). Figure 3.6 shows the market is concentrated upwards through

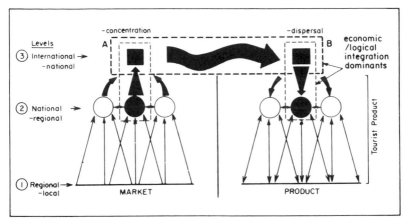

EXPLANATORY NOTE

Two separate structures, A and B link up through a bulk-carrying transport function, forming the functional mechanism. Each structure has **three levels** of operations, which partake in the **basic processes** – concentration and dispersal. In structure A the lowest functional level is regional-local, representing the market areas, from where potential tourist consumers are activated by level II marketing functions (travel bureaus, tour operators, travel clubs, etc.). The next upward movement, to level III, illustrates the bulking in preparation for the ultimate A-to-B geographic transfer. The dispersal process involves a structured descent within the tourist product, via the arrival mechanism (level III) to the resort-accommodation facilities (level II) where the dispersal tends to stabilize. The usually insignificant interaction with the local level, economic as well as cultural, is indicated by reciprocal arrows.

Horizontal interaction occurs on level II, and obviously on level III. For structure A this demonstrates the complementary role of operating service agents. In B the level II interaction illustrates an enclavic, institutionalized tourist circulation between resort facilities, a process often actively promoted by hotel management, or a certain type of resort hotel. This accentuated circulation within the level is at the expense of further direct dispersal of the tourist flow to the indigenous local level.

Horizontal and vertical economic/logistical integration dominants (dotted lines), link together structures, levels, and associated functions. Greytone symbols illustrate a wholly integrated system in terms of financing, management control, organization and dimensioning. International airlines – British Airways, Pan Am and T.W.A. – institutional industrial conglomerates, sometimes with vested interests in transportation – I.T.T., oil companies – typify the integrated sector of the mechanism.

Fig. 3.7   Hills and Lundgren's core–periphery model.
(*Source:* Hills and Lundgren, 1977).

the hierarchy, with the international transfer occurring between the dominant urban centres in the generating and receiving countries. Dispersal within the peripheral destination is more restricted, with the tourists moving from their point of arrival out to some resort enclave.

Earlier, Lundgren (1972) had described such relationships as an expression of 'a metropolitan economic hegemony par excellence'. According to Lundgren (pp. 86–7), these relationships: 'are basically a function of the technological and economic superiority of the travel-generating, metropolitan core areas as such and the willingness of the destination areas to adopt metropolitan values and solutions in order to meet the various demands of metropolitan travellers'. Noting the importance of the travel component at this scale, Lundgren stresses the dominant role of the metropolitan countries as air carriers who can, effectively and selectively, control the volume and direction of tourist traffic. The technical, economic and commercial characteristics of modern tourist travel favour the development of integrated enterprises, further reducing the possibility of local participation (Ch. 2). Moreover, metropolitan firms, whether they are airlines, tour operators or hotel chains, also have the advantage of direct contact with the tourist market, which they exploit fully with their commercial expertise (IUOTO, 1975).

A similar model of tourism in the South Pacific has been proposed by Britton (1980a, 1982) who sets his study in the context of dependency theory and stresses the broader political–economic structures in which tourism there has developed. Britton then goes on to quantify the extent of external domination, showing two-thirds of the total tourist receipts in Fiji went to foreign-owned companies in 1977, compared to one-half in the Cook Islands (1979) and 8 per cent in Tonga (1979). Moreover, a significant share went to expatriate Europeans: Fiji (14.5 per cent), Cook Islands (38 per cent) and Tonga (2 per cent). Britton (p. 352) argues 'there is a direct parallel between the extent of a country's historical structural conditioning as a colony and the presence of foreign capital'. However, changes in foreign/local participation in the tourist industry of any one country over time are not pursued in the manner of Bastin (1984) and Haider (1985). Britton also suggests (p. 346) 'that while all participants in the industry hierarchy profit to a degree, the overall direction of capital accumulation is *up* the hierarchy'.

The enclaves depicted in Fig. 3.6, emphasized by Britton in the South Pacific and observed by writers in other developing countries, e.g. Morocco (Berriane, 1978), Mexico (Cazes, 1980b), Bali (Jenkins, 1982) and the Caribbean (Pearce, 1987a), are seen not just as physical entities but also as representing social and economic structures. Goonatilake (1978, p. 7), discussing tourism in Sri Lanka, sees resort enclaves as 'islands of affluence within the country, walled in and separate from the rest of the population'. Some tentative findings by Pearce (1988b) in Vanuatu suggest that some 60 per cent of tourists' daylight hours there were spent in and around their hotel so that in behavioural terms the separation of tourists and residents can be very real.

Jenkins (1982, p. 239) identifies three basic characteristics of enclave development:

a)   The specific infrastructure is not intended to benefit directly the resident–indigenous community. Any 'spill-over' effect is purely gratuitous, emphasis is placed on the provision of infrastructure facilities for tourism development.

b) The site location is physically separate from an existing community of development, with the chosen site intended to generate its own transient tourism 'community'. The facility will be operated with minimum trading and social links with an existing community.

c) That the facility is most used, if not exclusively used, by foreign tourists. In this circumstance, a level of demands and services will be generated which the indigenous community could not afford to buy, even if it so wanted. In this case *structural enclavism* (i.e. a and b above) is reinforced by *price* enclavism.

Tourist enclaves develop for several often interrelated reasons. Firstly, as Jenkins notes, scale economies in infrastructural provision can be achieved by concentrating development in a limited number of areas. This can be a significant factor in any new area and in developing countries in particular where infrastructural deficiencies may be severe. Secondly, as Jenkins observes with respect to Tanzania and Rodenburg (1980) with regard to Nusa Dua in Bali, enclaves may be consciously promoted by planners and governments as a means of limiting contacts between residents and tourists in order to reduce the 'social and cultural' pollution of the latter. Thirdly, market forces draw developers to a limited number of sites with the concentration of tourist plant promoting a stronger image in more distant markets.

Tourist development in developing countries, particularly enclave development, is also often characterized by the role of international aid agencies in promoting tourism as a development strategy and funding basic infrastructural projects. Such aid may be sought by or channelled through government agencies or corporations, with the projects favoured frequently being localized large scale ones rather than dispersed small-scale enterprises. In the case of the Nusa Dua complex in Bali, for example, a Bali Tourist Development Corporation (BTDC) was created in 1973 'to be the agent in developing the site for a series of hotels; the hotels were to be built by private capital' (McTaggart, 1980, p. 461). The BTDC then sought and received a $16 million loan from the International Development Agency to undertake the necessary infrastructural development.

Similarly in the new Mexican complex of Cancun created *ex nihilo* on the Caribbean coast of Quintana Roo, almost half of the $47 million public investment in the first phase of development was financed by a loan from the Inter-American Development Bank. A further loan of $20 million in 1976 accounted for a similar proportion of the second stage of development (Cazes, 1980b).

Development of Cancun is the responsibility of FONATUR (Fondo Nacional de Fomento de Turismo), a central government agency whose aim is 'to achieve a controlled increase in tourism by

95

Plate 5 Coastal tourism on the Yucatan Peninsula: the Sheraton typifies many aspects of the large-scale, institutionalized enclave development at Cancun

Plate 6 Wanderlust tourism on the Yucatan Peninsula: much of the control of the 'Mayan circuit' has been in the hands of dominant Yucatecan families

improvement and expansion of existing resorts and the creation of new tourist zones focused on newly created cities' (Collins, 1979, p. 353). FONATUR has been responsible for the selection of the site (after an intensive nation-wide inventory of sites and resources), acquisition of the land, preparation of the plan, development of the infrastructure and servicing of the site prior to its sale to the private sector. FONATUR has also participated jointly in the actual development of various hotels but that has been left mainly to the private sector, with 80 per cent of the private investment estimated to be Mexican in origin. Hotels owned and operated by Mexican chains are complemented by hotels built by Mexican investors and managed by international chains. There is also some owner-operation by multinationals (Pearce, 1983).

Cancun has also been characterized by quite a strong domestic demand although this has declined in relative terms as the resort has developed (Pearce, 1983). The Mexican share of hotel guests fell from about three-quarters in 1975, when the resort opened, to just under half in 1980 as the total number of visitors increased from an estimated 100,000 to 460,000 in 1980. Cancun also experiences a favourable seasonality pattern due to this domestic/international mix with a reasonable constant pattern of demand being experienced throughout the year. Much of the domestic demand is in summer whereas a significant proportion of the international traffic occurs during the northern hemisphere winter.

Other areas

The development and ownership structure of Cancun contrasts quite markedly with that of the tourist industry in inland Yucatan where one of the few detailed studies of tourism in a non-coastal region of a developing country has been undertaken by Lee (1978). Prior to the development of Cancun, the region's tourism was centred on the 'Mayan circuit', based on the capital of Merida and the archaeological sites of Chichén Itzá and Uzmal, attractions which still remain important today. From her detailed ownership study, Lee concluded (p. 27):

> In examining the more important sectors of the tourist industry, it becomes clear that the largest enterprises are owned and/or managed by members of the upper class with a sprinkling of upwardly mobile middle-class individuals holding a number of the less important positions. Control over the major tourist services is further concentrated into the hands of a few families. This has been accomplished in two ways. First, the greater part are managed by the owners or members of their families. Consequently, control is more direct; that is, it is not delegated to hired management. Secondly, there are a small number of families that have consolidated their power through the acquisition of a string of enterprises in the different areas of the industry [hotels, travel agencies, car rental

agencies, restaurants, souvenir shops]. Under this arrangement they are able to operate more profitably because they enjoy something of a monopoly over the services that make up the 'tourist product'. In short, the important sectors of the tourist industry are not only controlled by the Yucatecan bourgeoisie but the major part of the concerns are owned by just a few families.

Lee (p. 31) attributes the emergence of a strong local bourgeoisie to three factors: the predominance of the hacienda as the key unit of production in the Yucatan; the domination of the economy by henequen which has the same base of wealth (land) as the traditional hacienda system; and Yucatan's geographic isolation and lack of communications with the rest of Mexico.

Lee did, however, observe some changes appearing in the structure of the region's tourist industry, notably through the granting of US car rental franchises and in the development of Cancun. The latter might be interpreted in terms of Stage III of Fig. 1.2, that is, the institutionalization of tourism. In the case of the Yucatan, such institutionalization takes on a definite spatial bias and is concerned with a particular type of tourism, that based on sun, sand and sea, more than on the cultural circuit.

In parts of Africa, coastal tourism development has been accompanied by other forms of tourism. In Kenya, the development during the 1970s of mass tourism along the coast, characterized by many of the features experienced by other developing countries, has been complemented by the expansion and intensification of safari tourism (Rajotte, 1987). Safari tourism has been encouraged by the State through the creation of more national parks and reserves which have not only been a burden on the taxpayer but also at the expense of traditional agriculture as Rajotte (p. 84) notes: 'For the peasant farmer or herder the costs of wildlife conservation are high and are experienced directly in terms of loss of potential forage and damage to crops, livestock, property, and life. Major benefits, however, accrue to private-sector investors, tour operators and the national government.'

Coastal tourism in Moroccan cities such as Agadir and enclave development along the country's Mediterranean coast, has been complemented by international wanderlust tourism to such inland centres as Marrakesh and Fez (Berriane, 1978). Smaller inland centres also attract a share of the growing domestic market though many Moroccan tourists do seek a holiday on the coast (Berriane, 1986). Much of the domestic demand is centred on more modest non-classified hotels and camping grounds which have a much higher degree of national ownership than the international class hotels. In 1981, Moroccan domestic demand was estimated to account for over a third of the total bednights spent in hotels and camping grounds, and more than half of all demand if stays with friends or relations or in rented accommodation are included.

In contrast, much of the expansion of tourism in Lesotho, Swaziland and Botswana has been due to the growth in gambling and pornography, with these small States constituting 'the "pleasure periphery" of the South African vice market' (Wellings and Crush, 1983, p. 207). The Holiday Inn in Maseru, built in 1970 and financed entirely by South African capital, accounted for 60 per cent of Lesotho's hotel capacity in 1973. Later another luxury hotel financed by domestic revenue and internal loans was opened and managed by Hilton International, with a second more modest domestically financed hotel also being managed by consultants. Lesotho has been heavily dependent on the South African market and has seen a decline in visitor numbers as the two major chains in South Africa have embarked on a programme of hotel and casino development within South Africa's nominally independent Bantustans. The hotel industry's interests there have been keenly encouraged by the South African government with Crush and Wellings noting elsewhere (1987, p. 107):

> For the state, this 'alliance' has had the advantage of giving considerable weight to its attempts to legitimise bantustan independence in the international forum and has also allowed it to replace a considerable portion of the costs of subsidizing the bantustan strategy with private capital. Tourism capital, has, for its part, discovered in the bantustans a highly profitable arena for investment.

Singh and Kaur (1985) have examined tourism in alpine areas in developing countries, namely in the Himalayas. Historically, Singh and Kaur identify three types of tourism in the Himalayas: religious tourism, bourgeois tourism and social tourism. The first involved Hindu pilgrimage to Himalayan shrines (p. 370): 'Austerity remaining the keynote, pilgrims' economy ensured local benefits and maximum participation of the people, enriching both the host and guest communities'. A second phase of development saw British expatriates establishing climatic resorts and sanatoriums throughout the middle Himalayas (e.g. at Simla and Darjeeling) during the first quarter of this century. These were later frequented by Indian élite. The term 'social tourism' is used by Singh and Kaur to refer to the more extensive flows of visitors, particularly trekkers, to the higher Himalayas which developed in the post-independence era (after 1947), especially as a result of the new strategic roads built after the Sino-Indian clashes. While the processes involved here are not described in great depth, some details are given of specific regions, for example Ladakh, where the number of tourists increased from 527 in 1974 to over 14,000 in 1982, 90 per cent of them being foreigners (p. 375):

> . . . in 1979 out of fifteen comparatively big hotels in Leh, ten were

managed by Ladakhis and remaining five were run by outsiders. Personnel in catering and other standard services were drawn from India, Kashmir and Nepal. Some Tibetan refugees also did share. Most of the smaller eating houses and *Dhabas*, were run by local people as a family business. Local farmers also changed to growing more vegetables to meet the tourist needs. Good sums were collected through the entrance fees to monuments and monasteries which the Gompa Association wisely spent on restoration and renovation of monasteries.

However, Singh and Kaur note (p. 375):

> Considered as a whole, only a small part of tourist earnings remains inside the Ladakh country, and much of it flows back to the Kashmir valley and the lowlands, affording a poor multiplier.

While tourist development in some of these areas, particularly Lesotho, follows a similar path to that observed in many insular situations in the Third World (the country is after all an enclave within South Africa), in other areas the process differs (e.g. in the Yucatan). More non-coastal studies in developing countries are now required to enable more general patterns to be identified.

Alternative tourism

Opponents of high-volume, large-scale, enclavic forms of tourist development have suggested that there must be other, better ways of developing tourism in Third World countries. Mass tourism, characterized by the features just cited, has been criticized for the degree of external control and for failing to deliver on promised economic benefits while causing severe social disruption (Britton, 1977; Rodenburg, 1980; O'Grady, 1981). These issues will be explored in more detail in Ch. 6. One consequence of this debate has been a call for the development of 'alternative tourism', though a general consensus on what constitutes 'alternative tourism' has yet to emerge.

After reviewing different interpretations of the term, Cazes (1986) suggests alternative tourism must be seen as a dynamic system consisting of three interrelated sets of components: values, processes and forms:

- the values underlying alternative tourism are based on the concepts of emancipation and self-determination and the search for spontaneity, enhanced interpersonal relations, creativity, authenticity, solidarity, and social and ecological harmony,
- the processes involved depend on a much fairer partnership between external and local entrepreneurs and organizers at all stages of development (initial research, legal and financial organization, construction, operation, management and sharing of benefits),

- the different forms (social, spatial, ecological, architectural) that alternative tourism takes must reflect the underlying principles of self-determination and integration: use of local materials and traditional forms of architecture, use of local builders, management and employees.

Similar characteristics are attributed to what Jenkins (1982) calls 'integrated development' as opposed to enclave development (and in contrast to the 'integrated' ski-resorts discussed in an earlier section and the vertical and horizontal integration reviewed in Ch. 2). The common features of 'integrated' (alternative) tourism he identifies are:

- a smaller unit scale of facilities,
- entry barriers are lower and there is more indigenous capital and management,
- because of the lower prices, the expectations of the guests attracted will differ from those staying in international class hotels. Such tourists may be more readily assimilated into the community and less socially disruptive,
- tourist acceptance by the host community may be less of a problem as such tourism emerges from rather than is imposed upon the community,
- 'price enclavism' may not be a feature given the smaller scale types of development related more to the local community.

These general perspectives on alternative tourism have been complemented by various case studies which not only provide some more detail on this form of tourism but also highlight some of the real world problems which have been encountered.

Ranck (1980) has examined different attempts to foster alternative tourism in some of the more isolated parts of Papua New Guinea since the early 1970s. In Tufi, this has taken the form of clan-run guest-houses catering for from six to eight people, built of local material by local people. In Goroka, the town council operates a couple of small lodges, one a converted convent. These operations appear to have met with mixed success; some of the more accessible guest-houses in the Tufi region have attracted several hundred guests per year while others have rarely had anyone overnighting. Problems have been experienced in terms of maintenance, water supply and especially with communication with potential guests or tour operators. But as Ranck points out (p. 62):

> Nonetheless, given the record of other heavily funded, heavily 'experted' development projects, the guest house phenomenon has been much more successful. As a small scale tourist business providing cash supplements to a subsistence society geographically isolated from other business communities, it seems well suited to the area. The local people are

proving competent managers, have maintained their dignity, autonomy, and are showing themselves eager to understand the wider world around them.

In Goroka, traditional native performances have been revived or organized for tourists, the best known being those by the Asaro valley mudmen. The mudmen have never had control of their operation; performances are negotiated with the promoter by a middleman. Communication between the various parties has again been a problem.

In French Polynesia, official emphasis has gone from luxury *hôtels d'impact* on Tahiti to bungalow-type accommodation on Moorea and Bora Bora to encouragement for small family units on the more distant islands, particularly in the Tuamotu archipelago (Blanchet, 1981). This latter policy was designed to maintain the population in these islands, but the small units developed with State subsidies have proved difficult to market and have experienced mixed fortunes. The first example examined by Blanchet, Te Anuanua on the Tahiti peninsula, was created by a rural development co-operative in 1968 and met with some initial success by attracting aircrew stopovers. However, it closed a decade later after this regular business was withdrawn and was unable to be replaced. Patamure Village, a score of small bungalows (*fare*) on one of the Tuamotu atolls, had an even shorter life. It closed in 1976 after only 2 years of operations, primarily as a result of the withdrawal of air access due to the inadequate standard of its landing strip. Tahaa Village near Raiatea in the Iles Sous le Vent, has been more successful but heavily dependent on a narrow market. Gradually developed since 1975 as a collection of bungalows by a local resident, Tahaa Village draws the bulk of its clientele from personnel on leave from the nuclear testing centre, the Centre d'Expérimentation du Pacifique. Difficulties of access and marketing have limited the owner's ability to expand his market to include more general visitors to French Polynesia. Blanchet suggests that investment aid alone is insufficient and that for the development of small hotels to be fostered successfully support must also be given in terms of training, management and especially marketing.

Similar initiatives and policies towards alternative tourism are also evident in the Caribbean (Pearce, 1987a). In the French *département d'outremer* of Guadeloupe, large hotels are concentrated along the Riviera Sud near Pointe-à-Pitre, but recent policy has stressed diversification in terms of both plant and location. Official encouragement and assistance is now being given to the development of small units, private and communal, initially on the Ile des Saintes and later on Marie-Galante and La Désiderade. The scale of these units, twenty rooms in the case of *gîtes communaux*, is to be in keeping with the small size of these islands. The emphasis is on water sports, walks and other activities rather than on the beach. Alternative accommodation on the island away from San Juan, notably villas and small *paradores*, has also been encouraged in Puerto Rico. St Vincent embarked on a policy of 'indigenous and integrated' tourism in 1972 based on gradual, small-scale, locally controlled development but the government lost power before it could be implemented fully (Britton, 1977).

Most of the 150 hotels in Belize are small, locally owned, family-run establishments, often built by enlargement of existing dwellings and offering modest standards of service and comfort (Pearce, 1984). Resort-style hotels, featuring cabanas and financed by foreign capital, are a recent development on Ambergris Cay. Most of the development there has been centred on San Pedro, a fishing village with 1100 inhabitants. The first hotels were built there in the late 1960s and by 1982 twenty-three hotels had a total capacity of 450 guests. In view of the gradual insertion of tourism into the community and the strength of the local fishing economy, San Pedranos have been able to participate in the new activity and to adapt successfully to the influx of visitors. The dominance of small-scale hotels in Belize is not so much a reflection of specific government policy as of limited resources and demand (a function in part of an inability to market the country's attractions). Any expansion is conditional on the provision of an adequate water supply, improvements in effluent disposal, upgraded access and joint promotion. In the case of Belize it is debatable whether the situation represents a form of alternative tourism or the initial pre-take-off stage of a more traditional form of development.

The most widely cited and apparently most successful of the officially sponsored alternative tourism programmes is that of the Lower Casamance region of Sénégal which is variously described as a 'tourism for discovery project' or 'integrated rural tourism' (Saglio, 1979, 1985). The programme was initiated in the early 1970s under the guidance of an ethnologist (Saglio) by the Agency for Cultural and Technical Cooperation in reaction to some of the impacts of the mass tourism experienced elsewhere in the country, for example the Cap Skirring Club Méditerranée. By 1983, nine small villages of traditional Diola dwellings had been built (a total capacity 310 beds) and 13,000 bednights per year were being recorded. Access is largely by canoe along waterways or the sea, the villages were built and are operated by the local residents with the distribution of profits being decided communally. However, financial assistance with construction costs was provided, for example by the Agence de Coopération Culturelle et Technique and by the Canadian Government.

Overall, the strategy followed appears to have brought together a number of positive features – low investment costs, modest prices, significant returns to the local community – and met a demand from a growing market segment interested in 'authenticity, contact and discovery'. Some difficulties with access have been experienced and an initial

Plate 7 Tourist development on San Pedro, Ambergris Cay, Belize: 'alternative tourism or the initial pre-takeoff stage of a more traditional form of development?'

period of adjustment was necessary, for example in handling the new influx of cash. Much of the success of the programme appears to be due to the Department of Tourism's role in co-ordinating and marketing this new and different form of tourism. Some initial demand was generated by way of short overnight excursions from the Club Méditerranée. Contacts were also developed with travel agencies specializing in youth travel. More recently, attention has turned to non-profit-making travel associations. Success in the Lower Casamance region has prompted ideas for further expansion in other regions of Sénégal and generated much interest from other developing countries.

The Casamance example does, however, raise questions about whether the alternative tourism projects there could have been established and survived without the prior growth in mass tourism and the accompanying markets and infrastructural support, for example the international air connections. In other words, can such low-key developments survive in developing countries without having some other form of tourism to be an 'alternative' to?

CONCLUSIONS

The typologies of tourist development presented in the preceding sections have demonstrated that tourism may develop in a variety of ways. While sets of common processes have been identified, considerable variation was shown in the types of tourist development occurring in developed countries and, to a lesser extent in developing ones. Much of this diversity is due to the variety of elements which might be developed and to the potential participation of a range of agents of development. The context in which development occurred was also shown to play a significant role in introducing variations on general themes as was well demonstrated in the cases of ski-field development in the European Alps and coastal tourism along the Costa Brava. Other variation might be attributed to the diversity of tourist demand, a point developed further in the following chapter. Moreover, if different processes give rise to different impacts then considerable diversity can be anticipated in the extent to which tourism contributes to development (Ch. 6).

The approach discussed in this chapter now needs to be extended to a broader range of countries, regions and resorts in order to test the wider applicability of the typologies presented and to derive new classifications of tourist development. In particular, North American researchers might consider more closely the adoption or adaptation of some of the criteria and methods used by their European counterparts and the enthusiasm for researching development processes on small islands might usefully be extended to other parts of the Third World.

While a firm base for work in coastal, alpine and rural areas has been established, more attention needs to be directed to urban areas and to other types of tourism not considered here, particularly national parks and other natural areas (Lundgren, 1982; Keller, 1987), wanderlust tourism, and thermal resorts.

Chapter

4

Demand and development

Demand and development are inextricably linked, though in a variety of ways. In general, the process of tourist development marks a continuing adjustment between supply and demand. As was noted in Ch. 3, with reference to Barbaza's (1970) study of the Mediterranean–Black Sea coastline, tourist development may be demand or supply led. In the first case, that of Spain, the provision of facilities and services oriented to tourists may arise as a response to growing market demand. In the second, as with the planned resorts of Languedoc–Roussilon and the Black Sea, development may be undertaken to stimulate demand. Whatever the initial impetus, successful development in the medium and long term necessitates a matching of supply and demand in terms of range, quality, quantity and price. An evolution on one side of the demand–supply equation will usually be accompanied by changes in the other, whether this represents growth, stagnation, decline (Fig. 1.4) or some qualitative change.

Moreover, the nature and extent of the demand and the associated facilities and services provided will also directly influence the broader aspects of development discussed in preceding chapters. The social, environmental and economic impacts of tourism, for example, are determined in part by the nature and level of demand. Indeed certain measures of demand, for example expenditure on tourist services and facilities, are also common (though simplified) measures of economic impact. The quest for alternative paths of tourist development discussed in Ch. 3 might also be seen as a consequence of disenchantment within some developing countries to catering for a certain type of demand, namely mass tourism, generated in foreign markets rather than meeting the needs of the local society. While a variety of alternative projects have been developed, experience has shown it has not always been easy to capture an alternative market and thereby successfully incorporate local residents into the development process. Where demand exceeds supply, serious externalities may arise as in the case of environmental pollution resulting from inadequate infrastructure.

An examination of demand is therefore important to an understanding of various facets of development. From a practical point of view, anyone engaging in developing services or facilities for tourists will require a reasonably detailed knowledge of their market and their customers' needs and wants. Analysis of demand through market research thus forms an integral part of the tourism planning process as discussed in Ch. 7. An understanding of demand is also essential for an appreciation of the broader development issues which tourism may generate. Impacts result not only from the provision of facilities and services but also the consumption *in situ* of these by tourists. These impacts in turn condition levels of social and economic development, however defined. Consequently, as Wahab (1975, pp. 92–3) points out, a new and broader perspective on tourist demand is needed:

> A quantitative approach of the tourist demand has usually been adopted by tourist destinations. The only pertinent question up to now, is how much does tourism yield to the country's economy ... yet, in view of the many sociological and economic problems that tourism might cause to the receiving country, it is necessary to start considering likewise a qualitative approach to the tourist demand asking ourselves, 'who is the tourist we want?'

This chapter seeks therefore to review elements of tourist demand and to discuss some of the developmental implications of these, leaving other issues for examination in later chapters. A general review of the nature of tourist demand and how it is measured precedes a discussion of two major components of demand, motivation and the ability to travel. Consideration is then given to more applied aspects of demand: decision-making, market segmentation and demand forecasting.

THE NATURE OF TOURIST DEMAND

Demand is one of the more complex aspects of tourism. It can be defined and measured in a variety of ways and at a range of scales. In standard economic terms, demand is generally taken to mean the quantity of a good or service that consumers, in this case tourists, are willing to buy at a specific price in a given period at a particular place. According to Wahab (1975, p. 91):

as a functional concept [demand] signifies the existence of a law of behaviour between several variables among which the nature of the product, its price and its utility figure most. Because of the difficulty involved in relating the volume of demand to so many variables at once, economists attempt to isolate what they consider . . . to be the most influential variable or few variables and to relate the demand volume to changes in them, assuming all other elements constant. In this way, for example, demand is usually expressed in relation to price.

While this might be accepted in certain commodities where prices are the most pervading factor, in tourism the functional relationship involved in demand is not that simple. The multitude of factors that intervene to induce consumers (tourists) to travel to certain destinations or to refrain from doing so operates in such a complex manner as to justify a different concept of demand.

In the case of coastal tourism, for example, major tourist activities such as swimming and sunbathing may incur no direct charges although certain indirect costs may be borne by the tourist, for example for travel and accommodation. Moreover, as P. Pearce (1982, p. 60) points out, the resources used might be valued in different ways: 'For one group the surf and sand may represent entertainment for children; for another group the coast provides a venue for social and possibly sexual encounters. This difference matters, both from a theoretical and marketing perspective'.

Wahab (1975, p. 92) expresses his 'different concept' of demand mathematically in the form of the following equation:

$$DAij = \frac{M.T.F.W}{R}$$

where

DA = actual demand
ij = from point (i) to destination (j)
M.T.F.W. = man–time–finance–will
R = resistance for various reasons amongst which are distance, costs, competition, political instability, bad image, lack of appropriate facilities, etc.

More generally (Pearce, 1987a), demand might be seen in terms of the relationship between individuals' desires or motivations to travel (which equates with Wahab's will) and their ability to do so (which is largely a function of time, finance and the resistance factors mentioned). These motivation and ability factors will be examined shortly but first it is useful to consider other definitional and measurement elements.

The actual or effective demand featured in Wahab's equation is generally used to refer to those who actually travel to a given destination or participate in a tourist activity. Effective demand is distinguished from deferred demand, that is 'those who could participate but do not, either through lack of knowledge, or lack of facilities or both', and potential demand which refers to 'those who cannot at present participate and require an improvement in their social and economic circumstances to do so' (Lavery, 1974, p. 23).

Effective demand is frequently measured in terms of numbers; the number of tourists leaving or visiting a country, region or city, using a certain mode of transport, taking part in a specific activity such as skiing or the number of some other unit, for example bednights spent in a particular type of accommodation or visits to national parks (Pearce, 1987a, pp. 113–127). For economists, effective demand is commonly expressed in dollars spent on a given activity in a particular locality or generated by a specific market.

Demand for tourism is also expressed in terms of travel propensity. Schmidhauser (1975, 1976) differentiates between two types of travel propensity, net and gross. Net travel propensity refers to the proportion of the total population or a particular group in the population who have made at least one trip away from home in the period in question (usually a year) and is calculated by the following formula:

$$\text{Net travel propensity (as \%)} = \frac{p \times 100}{P}$$

where

p = the number of persons in a country or in a particular population group who have made at least one trip away from home in the given period

P = total population of the country of group

Gross travel propensity refers to the total number of trips taken in relation to the total population studies and is expressed by the formula:

$$\text{Gross travel propensity (as \%)} = \frac{T_p \times 100}{P}$$

where

T_p = total number of trips undertaken by the population in question

P = total population of the country or group.

Schmidhauser also discusses the concept of travel frequency which refers to the average number of trips taken by a person participating in tourism in a given period according to the formula:

$$\text{Travel frequency} = \frac{T_p}{p} = \frac{\text{gross travel propensity}}{\text{net travel propensity}}$$

An increase in effective demand may reflect changes in net travel propensity, gross travel propensity or a combination of the two. Or, an apparently constant demand may conceal changes in the composition of that demand. Although the net travel propensity in the Netherlands fell from 62.6 per cent in 1981 to 60.2 per cent in 1982, the size of the Dutch holiday market remained at 11.7 million holidays due to an increase in travel frequency from 1.36 to 1.41 (Tideman, 1984).

The type of demand measure used is often simply a function of the statistics available for data deficiencies still continue to plague many areas of tourism analysis. In other instances, the measure of tourist demand used will depend on the nature of the problem being examined. Hollander (1982, p. 4) notes, for example: 'While numbers are important where capacity constraints are significant, such as in the design of airports and in the provision of aircraft, it is expenditure which affects income, employment and most other economic parameters'.

However, tourism, as has been stressed in earlier chapters, involves a composite experience made up of a number of different goods, services and activities. Thus the unit of demand may variously be the entire trip or vacation or various elements of it, for example use of a particular mode of transport or means of accommodation, which may be combined or segmented in an infinite variety of ways. Moreover, many units of demand can be measured at different scales and in a range of places from the origin to a series of destinations and back again. As an example of the great diversity of tourist demand which may exist, Oppedijk van Veen and Verhallen (1986) cite a Dutch survey of over 2000 holidaymakers of whom 40 per cent had unique vacations. The largest group having similar vacations (a two-person summer vacation, abroad, in the mountains, with hotel or apartment accommodation, transport by private car and no advance reservations) consisted of no more than 2 per cent of the sample.

As a consequence of its inherent diversity most research on tourist demand is far from comprehensive. Studies have been carried out both in origins and destinations and undertaken by researchers from different disciplines and for different purposes. Traditionally much of the work at destinations has been more applied and directed at ascertaining the economic determinants of demand and at segmenting different groups of users, often of specific tourist products. Such work has tended to emphasize the pull factors at the destination and those generating resistance to travel. Recent motivational research in the origins has underlined general push factors while other studies have been concerned with establishing general profiles of travellers and non-travellers. An attempt is made in the following section to draw these different strands together.

MOTIVATION

Motivation to travel or to participate in some form of tourism might be defined as 'that set of needs and attitudes which predisposes a person to act in a specific touristic goal-directed way. Motivation is thus an inner state which energizes channels and sustains human behaviour to achieve goals' (Pizam, Neumann and Reichel, 1979, p. 195). A distinction can be made between motivation and the more frequently reported purpose of visit or objective of travel. The distinction is illustrated by Pizam, Neumann and Reichel (p. 195):

> A tourist may be motivated to travel to attend a family function in order to satisfy any of his needs of belonging, status, or recognition, though his *stated objective* for such travel may be to visit friends and relatives. The difference between these two – motivation and objective – is that while the objective is a conscious and overt reason for acting in a certain way, motivation may be an unconscious or covert reason for doing it.

Understanding why people travel, why they engage in particular forms of vacation behaviour or select specific goods and services is fundamental to a full appreciation of tourist demand. While substantial progress has been made in this area in recent years much more work remains to be done. The focus here is on vacation travel though it must be remembered that there are other types of travellers, for example those travelling for business, education and health purposes. In many of these cases the critical demand factors are not so much those which generate the travel – as these in many respects are relatively fixed – but those which underlie the choice of specific travel products and services.

Many of the basic psychological factors underlying vacation motivation and behaviour were put forward in the pioneering paper by Grinstein (1955). Grinstein identified a need to 'get away from it all', a need to escape from the demands of everyday life which can best be achieved by a change of place. He concluded (p. 185):

> From the standpoint of the vacation itself, . . . there must be permitted a change to a type of milieu, varying with each individual, wherein the greatest possibility of increase in ego boundaries can result. Concomitantly with this, it is desirable that a greater opportunity exist for the development of a state wherein one feels at peace, wherein tensions pertaining to or developing from the individual's life situation are minimized, and wherein one is able to indulge one's pleasure principle to the maximum.

Later writers, while not drawing directly on Grinstein's work, have extended some of his ideas and developed some of the issues he raises.

Gray (1970) saw two basic reasons for pleasure travel – 'wanderlust' and 'sunlust'. Wanderlust he defines as 'that basic trait in human nature which causes some individuals to want to leave things with which they are familiar and to go and see at first hand different exciting cultures and places . . . The desire to travel may not be a permanent one, merely a desire to exchange temporarily the known workday things of home for something which is exotic'. Sunlust, on the other hand, 'depends upon the existence elsewhere of different or better amenities for a specific purpose than are available locally.' One expression of this, as the term suggests, is literally a 'hunt for the sun' as typified by tourist flows to the Mediterranean, Caribbean or South Pacific. Wanderlust might be thought of essentially as a 'push' factor whereas sunlust is largely a response to 'pull' factors elsewhere.

Reporting on a motivational study of thirty-nine individuals, Crompton (1979, p. 415) notes: 'The essence of "break from routine" was, in most cases, either locating in a different place, or changing the dominant social context from the work milieu, usually to that of the family group, or doing both of these things'. Having established a 'break from routine', as the basic motivation for tourist travel, Crompton then suggests that it is possible to identify more specific directive motives which serve (p. 415) to 'guide the tourist toward the selection of a particular type of vacation or destination in preference to all of the alternatives of which the tourist is aware. In most decisions more than one motive is operative'. Crompton goes on to suggest that his respondents' different motives can be conceptualized (p. 415) as being 'located along a cultural–socio-psychological disequilibrium continuum.' The socio-psychological motives were not expressed explicitly by the respondents but Crompton identified seven of these:

1. Escape from a perceived mundane environment.
2. Exploration and evaluation of self.
3. Relaxation.
4. Prestige.
5. Regression (less constrained behaviour).
6. Enhancement of kinship relationships.
7. Facilitation of social interaction.

Two primary cultural motives – novelty and education – were also expressed by Crompton's respondents. These would appear to be closely associated with some of the socio-psychological motives. The search for novelty, for example, might well be a complement to the escape from the mundane environment.

Crompton's work is supported by other studies. The regressive element (5) was highlighted by Grinstein (1955, p. 179) who noted: 'Lying on the warm sand, being buried in the sand, being practically in the nude are all examples of pleasures which in themselves represent manifestations of partial regression'. Crompton's seven socio-psychological motives also correspond well with the motivations identified by Dann (1977): reaction to anomie (1, 6, 7), ego-enhancement (4) and fantasy (4, 5).

Leiper (1984, p. 250) suggests, 'all leisure involves a temporary escape of some kind' but 'tourism is unique in that it involves real physical escape reflected in travelling to one or more destination regions where leisure experiences transpire'. He continues: 'A holiday trip allows changes that are multi-dimensional; place, paces, faces, lifestyles, behaviour, attitude. It allows a person temporary withdrawal from many of the environments affecting day to day existence.' As such, he argues, tourism enhances leisure opportunities, particularly for rest and relaxation. Leiper distinguishes between recreational leisure which restores and creative leisure which produces something new. He sees the three functions of recreation as being:

1. Rest (which provides recovery from physical or mental fatigue).
2. Relaxation (recovery from tension).
3. Entertainment (recovery from boredom).

Other writers have also portrayed vacation motivations in terms of the spatial inversion of normal routines or situations inherent in Leiper's functions of recreation. Gottlieb (1982, p. 167) argues that 'vacations *invert* an essential aspect of American socio-economic ideology and experience' and suggests that there are two basic cultural ideals for vacation styles held by Americans which she terms 'Peasant for a day' and 'Queen (King) for a day'. In the first, upper- or upper-middle-class Americans relax the social constraints of their home environment and on vacations abroad actively seek to mix with lower-class inhabitants of the host society and participate in events and activities whose equivalents back home they would studiously avoid. Conversely, on 'Queen for a day' vacations, lower-middle- to upper-middle-class Americans vacation to invert the social order by elevating their position in it. This may be reflected in behaviour where expenditure is more extravagant than normal, where attitudes of social superiority are exhibited, temporary interest in higher culture is shown and (p. 175) 'encounters with the lower-class natives are minimized for all but the "service" situations'.

This notion of ritual inversion in tourism has been developed and codified by Graburn (1983a) who expresses inversions as polar opposites on continua of a number of different dimensions (Table 4.1). Graburn emphasizes (p. 21) that 'tourists may choose the degree to which a particular behaviour . . . is a change from or reversal of their

Table 4.1 Inversions in tourism

Dimension	Continua (polarity and inversions, with examples)
Environment	Winter vs. summer: travel to the tropics and southern hemisphere Cold/darkness vs. warmth/sun: 'north–south' travel Crowds vs. isolation: wilderness, rurality; open space Modernity vs. history: often involving the 'opposite' urbanism Home vs. elsewhere: the 'trip' or reverse for itinerant workers
Class/ lifestyle	Thrift vs. self-indulgence: 'aristocratic' pretensions, gourmet cuisine Affluence vs. simplicity: slumming, mixing with 'the folks' Business-of-living vs. education: museums, special programs Superficiality vs. self-enlightenment: EST, etc.
'Civilization'	Urbanism vs. nature: 'Club Méd-type', beaches, wilderness Security vs. risk: climbing, rafting, trekking Fast pace vs. slow pace: avoiding distractions and demands Secular vs. sacred: religious pilgrimages, 'sacredness' of nature
Formality	Rigid daily schedule vs. flexibility: rising, going out, going to bed Rigid meal times vs. flexibility: snacks, etc. Formal/distant social relations vs. informality: *communitas* Formal clothing vs. informal/beachwear/nudity: sombreness, flamboyance Sexual restriction vs. licence: celibacy, marital, interracial
Health and person	Gluttony vs. diet: fat farms and camps Stress vs. tranquillity: mental relaxation or hobbies Sloth vs. exercise: sports, backpacking, walking Ageing vs. rejuvenation: baths, spas, cures Isolation vs. sociability: visiting or making friends: seeing family

(*Source*: Graburn, 1983a).

"normal" style and that each kind of tourism is characterized by the selection of *only a few key* reversals'. He also points out with regard to the inversions listed in Table 4.1 that these can be bidirectional, that polarities are often interrelated (e.g. familiar versus exotic and security versus risk) and that although tourists are rarely motivated by only one kind of behaviour reversal they only choose to switch a few aspects of their behaviour at any one time.

Many of these factors are brought together by Iso-Ahola (1982) who proposes a theoretical motivational model in which the escaping element is complemented or compounded by a seeking component (Fig. 4.1). One set of motivational forces derives from an individual's desire

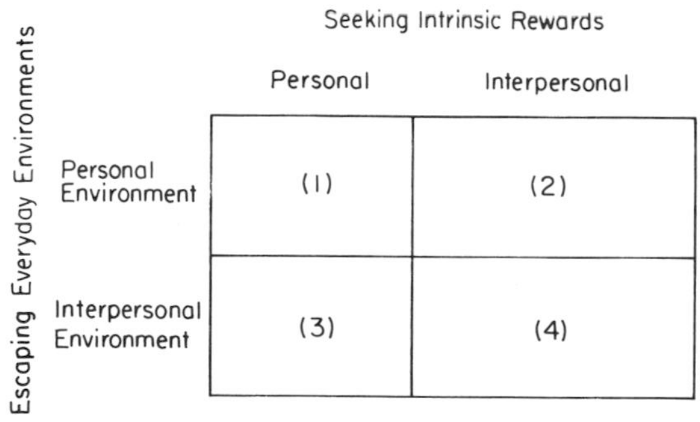

Seeking Intrinsic Rewards

Fig. 4.1 Iso-Ahola's social psychological model of tourism motivation. (*Source:* Iso-Ahola, 1982).

to escape his personal environment (i.e. personal troubles, problems, difficulties or failures) and/or the interpersonal environment (i.e. co-workers, family members, friends and neighbours). Another set of forces results from the desire to obtain certain psychological or intrinsic rewards, either personal or interpersonal, by travelling to a different environment. In general, Iso-Ahola's examples of personal rewards – rest and relaxation, ego-enhancement, learning about other cultures, etc. – and interpersonal rewards (essentially greater social interaction) correspond fairly closely with Crompton's specific motivational forces noted above and with the general factors raised by Grinstein. The point Iso-Ahola is making here, and which is conceptualized in his model, is that tourism 'provides an outlet for avoiding something *and* for simultaneously seeking something'. Recognition that elements of both sets of forces, whose relative importance may vary from case to case, may be satisfied at the same time is particularly useful in clarifying some of the issues that arise in motivational research. Iso-Ahola also suggests that in terms of dominant motives, it is theoretically possible to locate any tourist or group in any cell at a given time but notes that this cell may not only differ from one individual or group to another, but for any particular tourist or set of tourists it may change during the course of a trip or from one trip to another.

Evidence from large-scale empirical studies supports many of the points being made in these more theoretical papers. 'Change of environment', for example, was found to be the single most important factor accounting for the desire to travel abroad for vacation in a 1980 survey of 600 past and potential long-haul travellers in the United States (Opinion Research Corporation, 1980). A 1985 British survey of reasons for having a holiday abroad rather than in Britain identified better weather as the principal reason while in another survey that year 'relaxing' and 'a change of surroundings' were the first and second ranked factors sought by Germans on their main holiday (Marris, 1986). In each of these cases, as in most such surveys, a number of multiple responses were given, indicating that a variety of interrelated factors come into play.

The specific role of novelty in the travel experience was investigated by Bello and Etzel (1985) in a survey of Ohio and South Carolina consumers. Few differences were found between travellers to novel and commonplace destinations in terms of their demographics or the restful benefit of their trip. However, the novelty travellers felt their trip was more deserved and more educationally oriented.

Other empirical research has been concerned with more indirect measures of motivation. Woodside and Jacobs (1985) found the benefits derived from a visit to Hawaii varied from nationality to nationality. Rest and relaxation was reported as the major benefit by Canadians, mainland Americans emphasized cultural experiences while Japanese visitors stressed family togetherness. Pearce and Caltabiano (1983) analysed travellers' on-site positive and negative experiences in terms of Maslow's theory of a hierarchy of needs: physiological, safety, love and belongingness, self-esteem and self-actualization. They concluded that with positive experiences those studied found rewards in the fulfilment of physiological, love and belongingness and self-actualization needs whereas in negative experiences lower-order needs, especially safety, were more important. Time–budget studies of tourist activity patterns in Spain (Gaviria, 1975) and Vanuatu (Pearce, 1988b) raise questions about motivations in sunlust destinations with their studies suggesting only about 20 to 25 per cent of tourists' time there is actually spent at the beach or engaging in swimming and sunbathing.

What emerges from these different studies is that the fundamental motivation for tourist travel is a need, real or perceived, to break from routine and that for many this can best be achieved by a physical change of place. Thus 'change of place' is seen to be not just one of the defining attributes of tourism, but the very essence of it. Moreover, several writers suggest that the characteristics and conditions of the origin may be of equal or greater importance than the attributes of the destinations when analysing demand and planning tourist developments. Crompton (1979, p. 422) notes:

> The escape from a mundane environment, exploration of self, and regression motives, require only a destination which is physically and socially different from the residential environment. Literally thousands of destinations could meet these criteria and thus serve as direct substitutes . . . . The best promotion strategy for this market segment may be to stress the contribution the destination can make to reducing these disequilibrium states and to stress price advantages.

Similarly, Graburn (1983a, pp. 22–3) argues:

> the choice of tourist style stems from the culture and social structure of the home situation. . . . In the selection of changes that people wish to encounter in their tourism they choose those particular factors that they are *not* able to change in their home lives, within the constraints of opportunities offered by income and self-confidence.

The different sorts of escape and kinds of inversion outlined by Graburn give rise to different patterns of travel and types of activity and behaviour at the destinations (Table 4.1) with corresponding differences in development requirements and the impact of tourism. Earlier Gray (1970) had suggested that wanderlust and sunlust generate two distinctive forms of travel, with the former likely to be manifested more in international tourism in which travel remained an important ingredient throughout the vacation. Sunlust travel, on the other hand, tends to be more resort-based within a single country, often the traveller's own, with natural rather than cultural attributes being a key selection and development factor. Evidence of these differences is found in the distribution of international tourists in many European countries where long-haul tourists exhibit a preference for wanderlust destinations and short-haul tourists favour sunlust regions (Pearce, 1987a). In Spain, for example, demand from European visitors is highly concentrated in the Balearic and Canary Islands and the major Mediterranean coastal

Plate 8   The beach at S'Arenal: 'British tourists flock to the sunnier climes of Mallorca

provinces where charter tourism plays a significant role. Visitors from beyond Europe, on the other hand, focus primarily on the urban and historical centres which are visited by many as part of a cultural circuit.

Graburn's (1983a, p. 21) observation 'that each kind of tourism is characterized by the selection of *only a few key reversals*' is reflected in patterns of development in Spain's sunlust destinations. While many British tourists flock to the sunnier climes of Mallorca (Plate 8) they bring with them their own cultural baggage, often seeking their relaxation in British pubs and indulging their craving for traditional food at a local chippy (Plate 9). Likewise, in wanderlust destinations, the multinational hotel provides a safe and familiar base from which to venture briefly into an exotic culture. Alternative tourism, on the

Plate 9   'but they bring with them their own cultural baggage': The Chippy, S'Arenal, Mallorca

other hand, frequently requires not only cultural motives or interests but also a willingness to forego accustomed standards of comfort and convenience, an apparently less popular form of reversal. These examples illustrate some of the close links between tourist motivations and aspects of development in the destinations they visit.

THE ABILITY TO TRAVEL

To be able to satisfy their desire to travel, tourists must be able to meet various conditions. In particular, they must be able to afford both the time and money to do so and be able to overcome obstacles to travel.

Time and Timing

The growth of leisure time in the past century brought about by technological and other improvements has greatly increased the amount of time available for tourism, especially in Western societies. Many of the changes which have occurred have been reviewed in a recent WTO (1983a) study which also provides valuable information on current patterns of leisure time and the rights to holidays throughout the world.

The WTO study identified four basic types of time and quantified the average annual allocation of each for member countries of the European Community:

1. Biological time (43 per cent) – time required by individuals to satisfy all their natural needs or the time devoted to eating, sleeping and other biological functions.
2. Work time (34 per cent) – the span of time devoted to any remunerated activity.
3. Obligated time (7 per cent) – time given over to family obligations, social commitments, shopping, etc.
4. Free time (16 per cent) – that time remaining once people have been released from work and obligations and they have eaten and rested.

If time given over to sleep (34 per cent of the total) is excluded, the annual allocation of conscious time becomes: biological (15 per cent), work (48 per cent), obligated (13 per cent) and free time (24 per cent). Of the annual free time available, 6 per cent occurs on a daily basis, 11.5 per cent at the weekend and 6.5 per cent during holidays. In terms of tourism, it is these latter two periods, particularly the holiday time, that allows sufficient blocks of time for travel and stays away from home.

Both of these have increased significantly since 1960 with a widespread trend for a reduction in the working week and an extension of paid holidays (Table 4.2). Most of the countries where the average length of paid annual leave exceeds 24 days are industrialized nations while developing countries grant the shortest annual holidays. In Britain only 3 per cent of all full-time manual workers enjoying a basic holiday entitlement received 2 weeks paid holidays a year in 1951. By 1974 98 per cent were receiving 3 weeks or more. A decade later, 96 per cent of all full-time manual workers there received at least 4 weeks paid holidays (Papadopoulos, 1986).

Changes in the granting of annual paid holidays and the lengthening of the period of leave have clearly influenced the potential size of the tourist market in most countries. Moreover, the extensions of paid leave have often been accompanied by fragmentation of the leave taken. As a result, the pattern of vacationing in developed countries may be one in which main holidays are supplemented by shorter, secondary vacations. This trend has significant demand and development implications as the pattern of supplementary holidaytaking may differ markedly from that of main holidays as is well illustrated by a study of Belgian holidaymakers (Boerjan, 1984). Table 4.3 shows coastal tourism is much more dominant for Belgians' main holidays than for their supplementary ones when rural and other regions assume increased importance. Shorter supplementary holidays are often taken closer to home or used for specific purposes such as skiing.

The demand for tourism is influenced not only by the amount of leisure time available in given blocks but also the distribution of that time throughout the year. In most countries the demand for tourism is markedly seasonal in character. Factors such as institutionalized holiday periods when schools and many businesses close down or certain groups traditionally take their holidays and climatic conditions in the market areas and destinations give rise to 'peak' and 'off' seasons

Table 4.2 Trends in average length of leave with pay (Unit: number of countries answering survey)

Year	Number of days					Total number of countries
	<12	12–17	18–23	24–29	30+	
1938	4	7	3		2	16
1950	3	12	8	1	4	28
1960	3	16	16	3	6	44
1970	2	15	17	6	7	47
1982	1	13	17	14	15	60

(*Source* WTO, 1983a).

Table 4.3 Patterns of main and supplementary holidays in Belgium (1982)

Type of destination	Main holiday (%)	Supplementary holidays (%)
Coast	52.9	34.1
Mountains	18.6	18.7
Countryside	6.2	13.0
Forests/woods	6.8	12.3
Cities	4.7	7.9
Lakes/rivers	4.5	5.6
Ski resorts	1.9	7.3
Others	4.4	1.7

(*Source*: after Boerjan, 1984).

(Pearce and Grimmeau, 1985; Hartmann, 1986). The significance of the institutionalization of vacations is shown in a 1969 study from Belgium in which 61 per cent of those with no choice had their holidays in July compared with 34 per cent who were able to choose freely when they went away (Roucloux, 1977). The fragmentation of longer periods of leave noted earlier has contributed to some spreading of demand throughout the year, a trend reinforced in some countries by a conscious attempt to stagger holidays, particularly school holidays. Nevertheless, in many cases tourism still remains a very concentrated phenomenon.

The temporal concentration of tourism is shown clearly in Fig. 4.2 which depicts seasonal variations in hotel demand in Spain for selected nationalities and provinces (Pearce and Grimmeau, 1985). Most nationalities, including the Spanish themselves, follow, with some variations, the general pattern of a constant increase from January to the August summer peak followed by a steady decline through to December (Fig. 4.2a). The Benelux demand, for example, peaks in July while that of the French peaks more sharply in August, a reflection in both cases of established national vacation habits. In contrast, the demand from North America visitors is spread more evenly throughout the year, a function perhaps of the wanderlust tourism they practise being less influenced by climatic conditions than the European demand for seaside holidays. While the majority of the provinces exhibit a seasonal pattern comparable to the national average, two distinctive groups of major tourist provinces stand out in Fig. 4.2(b). The first, which includes the Balearic Islands and the north-eastern coastal provinces of Gerona and Tarragona, have a more accentuated summer season while the second (Madrid and the Canary Islands) is characterized by a remarkably constant level of demand the year round. The absence of marked

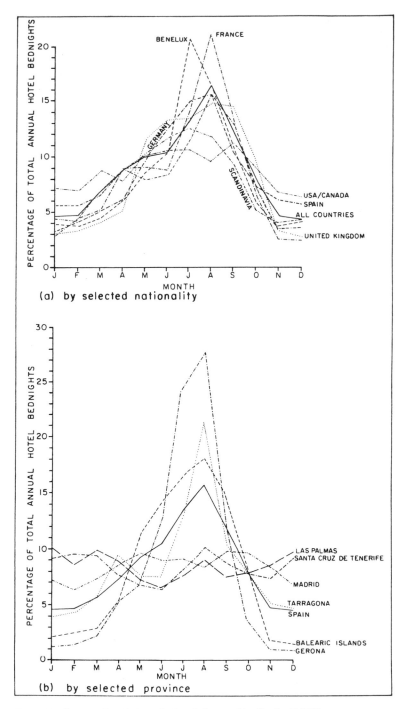

Fig. 4.2 Seasonal variations in hotel demand in Spain (1980).
(*Source:* rcdrawn from Pearce and Grimmeau, 1985).

peaks and troughs in Madrid results from its wide range of urban and cultural functions which are less seasonal in nature and from a market mix which includes a large share of the North American demand. The more equable climate of the more southerly Canary Islands means Santa Cruz and Las Palmas appeal to sunseekers the year round, including a significant number of Scandinavians during their harsh winter months. In contrast, Gerona's Costa Brava has a much ruder winter which, when coupled with the importance there of the French market in summer, gives it a sharp July–August peak.

The pronounced seasonality of provinces such as Gerona produces a variety of consequences for tourists, developers and host societies. For the tourist, seasonal concentration of demand often results in overcrowding, lack of choice and inflated prices. All of these may reduce the benefits of the holiday though the more gregarious may find the increased hustle and bustle more exhilarating than the calm of the off-season. For developers or operators, seasonality may mean a less than optimal utilization of plant, reduced profitability, difficulties in obtaining staff and supplies and cash-flow problems in the off-season. For these reasons a more balanced pattern of demand such as that shown for Madrid is often sought. Acute seasonal demand can often exacerbate the social, economic and environmental problems experienced by host societies and ecosystems, a point developed in Chs. 5 and 6.

Income, prices and elasticity

To be able to travel, tourists must not only have the time but also sufficient financial resources to enable them to do so. The economic determinants of tourist demand are basically income and prices (Wahab, 1975; Sauran, 1978; Hollander, 1982; McDermott and Jackson, 1985; McIntosh and Goeldner, 1986).

The increase in paid holidays noted earlier has generally been accompanied by rising real incomes which have meant more people have been able to afford to travel. Spending on tourism is usually regarded as discretionary spending for although travel has been expressed earlier as a response to a desire or need to escape, in an overall hierarchy of needs this usually comes after more basic ones such as food, housing and clothing have been satisfied. Even in some developed Western societies not all groups will yet have such discretionary power as shown by Meunier's (1985) study of low-income families in Quebec.

The tourist's economic ability to travel and engage in various forms of tourism depends not only on the level of income remaining after these needs have been met but also on the price of the different components of the tourist experience. In aggregate econometric analyses of international tourist demand, a distinction is usually made between the cost of travel (especially air fares) and destination or ground costs. For international travel, the exchange rates between the currencies of the market and destination will also affect the tourist's purchasing power. Technological and organizational factors, for example the development of wide-bodied jets used in charter tour operations, have played a significant role in the growth of mass tourism by bringing down the costs of tourism and making it more accessible to a larger proportion of the population.

Economists express the relationships between tourist demand and these different variables in terms of elasticity, that is, 'the degree of demand, responsiveness to changes in price structures or changes in various economic conditions of the market' (Wahab, 1975, p. 93). The measure used is the coefficient of elasticity which measures changes in demand (e.g. number of arrivals, bednights, etc.) resulting from some change in the determinant (e.g. income, air fares, ground costs):

> Specifically, an elasticity coefficient indicates the percentage shift (positive or negative) which can be expected in the number of arrivals as a result of a percentage shift in the level of the determinant. For example, a price elasticity of -1.0 implies a one per cent reduction in the number of arrivals for every percentage increase in the measure of prices. An income elasticity of 2.0 indicates a two per cent gain in arrivals for every one per cent increase in incomes (McDermott and Jackson, 1985, p. 17).

The elasticity coefficient can be obtained from the formula:

$$\text{Elasticity coefficient} = \frac{\%\ \text{change in demand}}{\%\ \text{change in determinant}}$$

While Wahab (1975) considered tourist demand to be 'highly elastic', that is it had a high coefficient value, subsequent analyses, surveys and reviews suggest a less straightforward picture. In their wide-ranging reviews, Sauran (1978) and Sunday (1978) identify income as the most significant economic variable explaining demand with varying degrees of responsiveness being reported for travel costs and relative prices. The significance of different economic determinants will of course vary depending on the markets, destinations and types of demand being considered. Hollander (1982), in an econometric analysis, found air travel by short-term foreign visitors on pleasure trips to Australia to be highly responsive to changes in personal incomes in the country of origin but not very responsive to changes in fares or relative prices in Australia and alternative destinations. Gunadhi and Boey (1986, p. 248), concluded that demand for tourism in Singapore 'is invariably highly income elastic, whereas the effects of prices and exchange rate movements vary from country to country'. In the case of foreign travel to the United States, Loeb (1982, p. 18) found that income, exchange rates and relative prices 'proved to have a significant effect on the demand for travel in the USA. However, the degree of responsiveness of the demand for travel attributed to the various variables varied from country to country.'

The determinants of demand and their elasticities may change over time as Guitart (1982, p. 24) notes: 'With rising incomes, there come changes in consumption patterns, reflected in different allocations to the main items on the household budgets, and in new needs. Thus paid holidays have led to tourism becoming an item on the household budget.' One consequence of this may be that the demand for tourism becomes inelastic, with expenditure on tourism being maintained when incomes stabilize or fall through a reallocation of spending on other items. Guitart suggests this has already happened in the United Kingdom.

Other evidence for such a trend comes from Belgium where Boerjan and Vanhove (1984) report savings on vacations by holidaymakers during the recession of the early 1980s were less than on heating and automobiles. Their study of Belgian and West German holidays over the period 1980–82 showed that although economic conditions did not have a great influence on the net propensity to take holidays ('Taking a holiday is for many people "a must"') and the average length of stay declined only marginally, significant changes did occur in the forms of tourism practised. Belgian and West German holidaymakers reacted to the crisis by taking more domestic and fewer foreign holidays, by reducing travel by air and by using more self-catering and less hotel and full-board accommodation. For example, use of second homes increased as owners spent more time in their own residence and as they also sought to economize by letting their second home out to other tourists. This substitution of demand is characteristic of tourism given the diversity of its component parts and the range of destinations available.

Such behaviour indicates that the tourist acts as '*homo economicus*', an assumption made in many economic demand studies. However, Guitart (1982, p. 38), in interpreting the mixed results of his analysis of the British package-tour market, cautions against the belief that tourists consistently engage in rational economic behaviour: '*homo peregrinus*, unlike *homo economicus*, permits himself sometimes (if not often, and sometimes of his own accord and sometimes unwillingly) to break the rules of economically rational behaviour'. Indeed, the inversions discussed in the motivations section, notably 'Queen for a day' and 'Peasant for a day', suggest that the tourist's economic behaviour while on holiday will specifically differ from its normal pattern.

Obstacles to tourism

The demand for tourism generated by increased time and disposable income is still tempered by a wide variety of obstacles which may act as barriers to individual travel or limit the aspirations of tourist developers.

The friction of distance is a limiting factor common to both domestic and international tourism (Pearce, 1987a). The basic geographic concept of distance decay, whereby the volume of tourist traffic decreases with distance away from the generating area, is embodied in several models of tourist travel, whether concerned primarily with domestic (Greer and Wall, 1979) or international (Yokeno, 1974; Miossec, 1976, 1977) tourism. Factors accounting for this decline with distance include increasing costs in time, money and effort, together with decreasing awareness of opportunities available. However, it is recognized that regular distance decay curves may be modified in the real world by 'positive deformations' (low cost of living, favourable climate, historic links) or 'negative' ones (essentially political). Empirical studies confirm the existence of a distance decay function in much domestic tourist travel, with variations being found amongst different groups of tourists. A broad distance decay effect can also be observed in international tourist flows, with intraregional traffic predominating in many parts of the globe and reciprocal flows developing between countries, particularly neighbouring ones.

Other impediments to tourism, especially international tourism, are government imposed (Ascher, 1984). For the wide variety of reasons outlined in Ch. 2, governments may seek to control both inbound and outbound tourists through passport and visa requirements, foreign exchange controls, custom duties and other regulations. Passports and exit visas are not automatically available everywhere, for instance in Eastern Europe; nor is freedom of entry universal. According to Ascher, more than 100 countries have restrictions limiting the amount of currency their citizens can obtain for foreign travel. Part of the worldwide upsurge in Japanese tourists in the 1970s can be attributed directly to the liberalization of Japanese currency exchange regulations in 1964 and to the easing of procedures to obtain passports in 1970 (Pearce, 1987a, pp. 55–60). Equally important for the development of tourism are the controls on tourism companies and businesses. The most common of these include: restrictions on foreign remittances, local equity requirements, operational limitations (restrictions on advertising, hiring non-local personnel) and control of access (particularly for airlines).

Lack of development, for these or other reasons, will also limit demand for destinations must be physically accessible to the tourist. This implies at least a minimum of demand and infrastructure. Political instability is also a major handicap for uncertainty and personal insecurity are scarcely compatible with the quest for rest and relaxation! Sharp downturns in the tourist traffic have thus accompanied political upheaval, war and terrorist activity in different parts of the world (e.g. Greece, late 1960s; Spain, 1974 and 1976; Fiji, 1987). However, tourist memories have been shown to be relatively short, with a return to business as usual with a resumption of normal conditions.

Tourists must also generally be in reasonable health and be able to travel. Age, illness and disability may be significant though not insuperable barriers to travel (Woodside and Etzel, 1980). Young children may be a further constraint on travel. Tourists must also be aware of holiday destinations and have the ability to plan and organize their trips.

PATTERNS OF DEMAND

The various demand factors reviewed in preceding sections combine in different ways to generate a range of patterns of demand. Selected examples will be examined here to illustrate different scales and types of demand, the factors which generate them and some of the resultant implications for development. These examples will also provide some further background to the various processes of tourist development discussed earlier. Attention will be directed initially at the evolution of demand, first at a global scale, then at a regional level with respect to the South Pacific. Socio-demographic profiles of European tourists will then be presented, followed by a case study of a particular type of accommodation, namely motels.

The evolution of demand

Figure 4.3 depicts the evolution of international tourist arrivals throughout the world and by major region from 1960 through to 1986. While the WTO figures on international arrivals are subject to some limitations (Pearce, 1987a, pp. 35–7), they nevertheless provide a good general picture of the global evolution of international tourism. With the exception of a slight check in 1968, Fig. 4.3 shows a strong and reasonably consistent growth in arrivals through to the early 1970s. Factors accounting for this growth include absolute population increases, rising standards of living, technological changes, and extensive and ambitious tourist development programmes in many countries by both the public and private sectors. A levelling off in demand occurred in the mid-1970s as national economies and individuals adjusted to the consequences of the energy crisis, including a quadrupling of aviation fuel prices between March 1973 and 1974 (Cleverdon and Edwards, 1982). The rate of growth picked up again later in the decade only to fall again in the early 1980s as the effects of the world recession were felt. In 1982 arrivals are estimated to have declined by 0.7 per cent over the previous year, the first absolute decrease recorded over the

Evolution of International Tourism 1960-1986

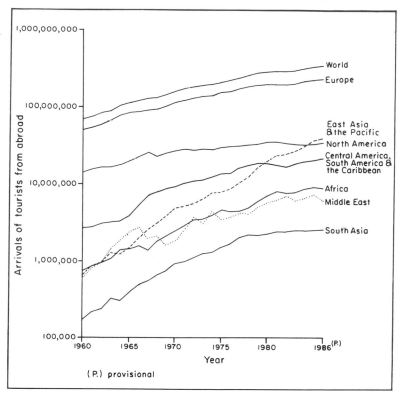

Fig. 4.3 Evolution of international tourist arrivals (1960–86).
(*Data Source:* WTO).

period shown. An upturn has been experienced in subsequent years (1984, up 7.3 per cent over the previous year), though the provisional 1986 increase in worldwide arrivals was only 2.6 per cent. Despite this slowing in the relative rates of increase, total arrivals are estimated to have grown 50 per cent over the decade 1975–85, passing from 214 million to a provisional 325 million in 1985. According to Lee (1987, p. 3), 'Global tourism has clearly proved its vitality and resistance to inflationary and currency pressures, political instability, unemployment and limitations on purchasing power.'

Regional variations are also apparent in Fig. 4.3. The growth of visitors to Europe closely parallels the worldwide trend as the region has retained its dominant position throughout the period (73 per cent of all arrivals in 1960; 67 per cent in 1985). Some of Europe's dominance can be attributed to definitional and geographical factors – 'international' travel is recorded much more readily there than in the USA given the small size and continental location of European countries – but it is also due to high standards of living in these relatively densely populated, industrialized countries. International arrivals in North America have not followed the same pattern of consistent increases, with absolute declines in traffic being experienced earlier at times of economic recession (1968, 1972, 1974) and for three successive years in the early 1980s (1982, -5 per cent; 1983, -3.4 per cent; 1984, -1.8 per cent). As a consequence, the region's share of world arrivals has halved over the period (20 per cent in 1960; 10 per cent in 1985), with North America now replaced by East Asia and the Pacific as the second most popular destination (11 per cent of arrivals in 1985). Part of the explanation for the extraordinary growth in this region is found in the rapid expansion of Japanese travel abroad, especially after 1970 (Pearce, 1987a). Traffic to Central and South America and the Caribbean developed rapidly in the late 1960s, then grew more slowly through to 1980 when the downturn in arrivals preceded the global trend by 2 years. Africa and the Middle East have seen greater fluctuations in their arrivals but the overall trend is one of growth, even if their total share of the world market remains under 3 per cent each. After almost two decades of strong growth from a small base, the international traffic to South Asia levelled off in the late 1970s and the region continues to record less than 1 per cent of the world's arrivals.

Explanations for the global evolution of international tourism shown in Fig. 4.3 must necessarily be of a general nature. At a smaller scale, more specific factors influencing the evolution of demand can be identified. Such is the case with the growth of international tourism in the South Pacific where arrivals have not increased at the same rate as the broader East Asia and Pacific region. While variations do occur from destination to destination, Fig. 4.4 shows a general pattern of slow growth and a levelling off in demand in the late 1970s and early 1980s. The similarity of the trends depicted in Fig. 4.4 suggests that tourism in the region has been influenced by factors shared by many of the destinations as well as by specific events which have had an impact on individual countries in particular years. Three sets of factors might be identified – those in the markets, those in the destinations and those determining the links between the two.

The two major markets for the region – Australia and New Zealand – have behaved rather differently throughout this period. At 7.2 per cent per annum for the decade and 6.1 per cent for the years 1979–84, Australian departures for the South Pacific increased at a greater rate than the Australian outbound traffic as a whole (respectively 6.1 per cent and 3.7 per cent). This was part of a more general growth in departures for short-haul tropical destinations at the expense of New

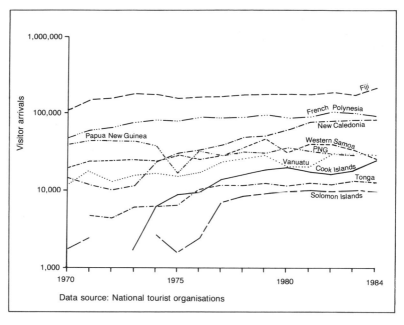

Data source: National tourist organisations

Fig. 4.4 Evolution of visitor arrivals in the South Pacific (1970–84). (*Data Source:* National tourist organizations).

Zealand and long-haul European destinations. In contrast, total New Zealand departures for the period 1979–84 decreased by an average of -2.1 per cent per annum as rising inflation and declining exchange rates took their toll. This decline was compounded by the erosion of the South Pacific's share of that country's market, which fell from 23.3 per cent in 1977 to 13.6 per cent in 1984. As a consequence of these trends, destinations in Melanesia which draw heavily on the Australian market have shown higher rates of growth (New Caledonia, Vanuatu) than those in Polynesia which have traditionally depended on New Zealand residents (Cook Islands, Western Samoa, Tonga).

Specific internal events have produced major fluctuations in the tourist traffic to particular destinations. Cyclones, for example, checked the expansion of arrivals in both Fiji and French Polynesia in 1983. In the latter, the effect of the cyclones was compounded by industrial unrest in the hotel sector and the temporary closure of the Club Méditerranée on Moorea for refurbishing in 1984. Political events have also left their mark, as in the downturn in visitor arrivals associated with independence in Papua New Guinea (1975) and Vanuatu (1980), where secessionist problems carried internal disturbances and a dampened demand through to 1981. More generally, several of the Pacific Island governments, such as that of Vanuatu, have adopted a cautious attitude towards tourism, with only limited resources being made available for tourist development and promotion (see Ch. 7).

The links between the markets and destinations which are essential for the development of international tourism in all parts of the world are particularly crucial for the South Pacific States characterized by their insular nature and remoteness from their major markets. Consequently the nature and capacity of air services has had a marked impact on the volume and composition of demand. For example the rapid growth in visitor arrivals to the Cook Islands in the mid-1970s followed the opening of the international airport on Raratonga in 1974 while the subsequent drop in visitor numbers in the early 1980s is partially attributable to revised schedules and reduced capacity. The upturn in visitor arrivals to French Polynesia in 1982 is in large part due to the reinstatement of the QANTAS Sydney–Papeete–Los Angeles service in September 1981. Conversely, Fiji lost a considerable volume of stopover flights from Australia and New Zealand to North America with overflying contributing significantly to the slow growth in arrivals shown in Fig. 4.4.

Tourist demand in the European Community

A comprehensive survey of tourist demand in what is perhaps the world's largest tourist market, the European Community, was undertaken in 1986 (European Omnibus Survey). This produced detailed data on holidaymaking for the previous year and is particularly valuable not only because of the cross-country comparisons but also because it includes details of those who did not take a holiday as well as those who did. It is based on a representative sample of almost 12,000 respondents aged 15 years and over. Holidaymakers are defined as people taking a holiday for more than 4 days away from their residence.

For the Community as a whole, an estimated 56 per cent of the adult population took a holiday away from home in 1985. This amounted to some 145 million adults or a total of 185 million holidaymakers when children are included. The propensity to take a holiday varied from about two-thirds in the Netherlands and Denmark to about one-third in Portugal (Fig. 4.5).

In the survey, non-holidaymakers were divided into two comparatively even categories, regular non-holidaymakers (those who in addition to not being away in 1985 had not taken holidays in 1984 nor intended doing so in 1986) and others (those who had not been away in 1985 but had taken holidays in 1984 and/or intended to go away in 1986). Holidaymakers in 1985 were categorized into those who had taken holidays once (37 per cent of the total) and those

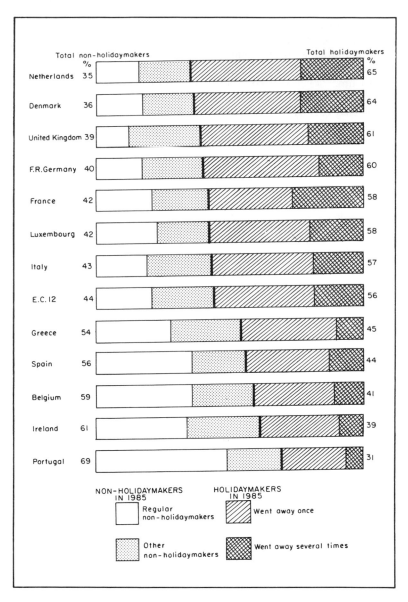

Fig. 4.5   National variations in holidaymaking in the European Community
(1985).
(*Source:* redrawn from European Omnibus Survey, 1986).

who had been away two or more times (19 per cent). In these terms, the countries of the Community can be divided into two groups. In the first, the more industrialized nations (the Netherlands, Denmark, UK, FR Germany, France, Luxemburg and Italy), the majority of the inhabitants take an annual holiday (and in many cases several) and the proportion of regular non-holidaymakers is less than 25 per cent. With the exception of Belgium (for which other surveys give a higher travel propensity; Boerjan, 1984), the second group is comprised of four more peripheral and less developed countries (Greece, Spain, Ireland and Portugal). In these, only a minority take holidays, relatively few go away more than once, and regular non-holidaymakers contribute the largest share of those not taking holidays.

Results from the EC survey suggest that while about a fifth of the non-holidaymakers preferred to stay at home, others wanted to get away but were constrained in some way from doing so. Economic factors were the major constraint (44 per cent), followed by work ties (16 per cent), exceptional family circumstances, poor health, moving residence and unforeseen impediments. The importance of these factors varied from country to country. In Denmark, for example, 38 per cent said they preferred to stay at home while in the peripheral countries the majority of non-holidaymakers could not afford to leave home (Portugal, 67 per cent; Ireland, 61 per cent; Greece, 55 per cent).

The impact of economic factors on demand is also reinforced by the socio-demographic profiles of holidaymakers and non-holidaymakers which show the most significant discriminating factors are two interrelated economic ones, occupation and income (Table 4.4). The propensity for holidaymaking, and especially multiple departures, is much higher in the upper occupation and income categories while regular non-holidaymakers are much more numerous in the low-income groups and amongst farmers and fishermen. Age also has some influence, with the propensity to holiday being lowest amongst the elderly. The presence of children, however, does not appear to play a major role, except in families with three or more children.

Less immediately apparent is the relationship between the holiday propensity and the size of urban area and type of dwelling. Rates of holidaytaking are shown to increase with size of urban area and with multi-unit dwellings. Why this should be is not altogether clear. It may be that average incomes are greater in larger cities or that rural populations are more elderly. However, other more detailed studies from France (Baretje and Defert, 1972) and Belgium (Boerjan, 1984) show that the rate of holidaytaking increases with city size even when incomes are held constant. This may reflect then motivational factors, with the need to escape increasing with city size or it may simply indicate that larger cities have better transport linkages and that it is easier to get away on holiday.

Table 4.4 Profiles of holidaymakers and non-holidaytakers in the European Community (1985)

	Did not go away in 1985			Went away in 1985		
	Total	Don't usually go away	Others	Once	Several times	Total
	(%)	(%)	(%)	(%)	(%)	(%)
Total EC 12	44	21	23	37	19	56
Income						
High R++	25	7	18	43	32	75
R+	40	15	25	39	21	60
R−	51	27	24	37	12	49
Low R−	64	41	23	27	9	36
Occupation of head of household						
Executives, top management	15	2	13	42	43	85
Professionals	18	4	14	39	43	82
White collar	29	8	21	44	27	71
Business, shop owners	44	19	25	40	16	56
Manual workers	49	22	27	39	12	51
Retired	51	31	20	31	18	49
Others, non-active	56	33	23	30	14	44
Farmers, fishermen	75	51	24	20	5	25
Age						
15–24	38	11	27	41	21	62
25–39	38	14	24	41	21	62
40–55	47	24	23	37	16	53
Over 55	53	33	20	30	17	47
Number of children (<15 years) at home						
None	44	23	21	37	19	56
One	44	18	26	38	18	56
Two	40	15	25	41	19	60
Three or more	56	28	28	30	14	44
Urbanization						
Large town	34	15	19	42	24	66
Small town	41	18	23	38	21	59
Village	55	30	25	32	13	45

Table 4.4 (Contd)

	Did not go away in 1985			Went away in 1985		
	Total	Don't usually go away	Others	Once	Several times	Total
	(%)	(%)	(%)	(%)	(%)	(%)
Type of dwelling						
Flat in block of						
over 50 flats	34	12	22	39	27	66
11 to 50 flats	35	15	20	44	21	65
up to 10	40	18	22	38	22	60
Semi-detached or terrace						
house	47	22	25	37	16	53
Detached house	46	23	23	35	19	54
Farm or isolated country						
house	62	41	21	29	9	38

(*Source*: European Omnibus Survey, 1986).

This inability to interpret readily certain of these factors does point to some of the limitations of what are valuable but essentially descriptive surveys. For this reason, other approaches to segmenting tourist demand have been undertaken, as is discussed later.

Motel location and choice

The examples given so far have dealt with fairly gross patterns of demand. These are essential in identifying general trends and interpreting these for planning and development purposes. Ultimately, however, tourism demand is reflected in the use of specific facilities and services at the local or sectoral level and it is important to analyse how and why this final 'consumption' of the tourist product takes place. While a number of marketing studies have considered accommodation selection processes (Mayo, 1974; Lewis, 1985), these have essentially been aspatial in nature and have not taken account of intra-urban variations in demand. The example discussed here, motel location and choice in Christchurch, New Zealand (Pearce, 1987d), combines the spatial analysis of the urban morphological studies discussed in Ch. 3 with the segmentation and choice aspects of marketing to be examined in the following section. Specifically, the study sought, through a questionnaire survey in November and January, to establish differences in the profiles of visitors to three groups of motels – those in the central city, on the major access routes and in the suburbs – and to examine the factors affecting the choice of motels in each of these three zones.

Analysis of intra-urban variations in the demand for Christchurch motels revealed significant variations in the profiles of motel guests and their choice of a motel in a particular part of the city. Key differences emerged for such factors as purpose of visit and the main reason for selecting a specific motel but for other variables such as length of stay and the extent of pre-booking no significant variation was found. The variables on which major differences were found are depicted in Tables 4.5 and 4.6. Overall, the sharpest contrast occurred between the central city and suburban motels, with the major route motels constituting something of an intermediate category between the two. Some seasonal variation was also apparent between the November sample (a 'typical' week) and the January one (a 'peak' week) but for space reasons only the November results are shown in Tables 4.5 and 4.6.

The extent to which the present pattern of motels will satisfy future demand will in large part depend on the degree to which the market for motels remains constant or evolves, a factor to be taken into account by city planners responsible for granting building permits. Changes in the domestic demand, either an increase or a downturn, will most likely be felt more in suburban areas and to some extent on the major routes. Expansion or contraction of the overseas market, and of the holiday market in general, will have a greater impact on central city motels. More attention might be paid at the local level to the significance of the motel sector to the overseas visitor, many of whom find the service and facilities offered by the motel particularly appealing. Individual moteliers might use the profile and decision-making findings of this study to direct their marketing effort more effectively.

DECISION-MAKING

The preceding sections have shown the demand for tourism to be characterized by a diversity of motivations, differing abilities to travel and different patterns of demand. Each of these factors can influence the level and type of tourist development and in turn be influenced by that development. Understanding the interrelationships between the different facets of demand and the interaction between demand and supply is an integral part of tourist development. A useful first step in developing this understanding and matching demand and supply is to consider the tourists' decision-making or consumer behaviour.

Hodgson (1983) suggests that questions affecting the decision to take a holiday are hierarchical in nature (Table 4.7) with higher order questions largely being predetermined and less subject to influence by marketing and development. Rather, he believes (p. 19): 'the answers to these questions form a sort of framework within which

Table 4.5 Profiles of Christchurch motel guests by location (November, 1985)

	Central city (%)	Major routes (%)	Suburbs (%)	All (%)
Purpose of visit				
Vacation	50.0	33.0	34.2	39.9
Business	27.1	40.6	21.7	29.2
Visiting friends/relations	11.4	5.7	17.5	11.7
Other	11.4	20.8	26.7	19.1
Day of arrival				
Monday	19.9	13.1	12.0	15.3
Tuesday	15.6	18.7	6.8	13.7
Wednesday	14.9	16.8	16.2	15.9
Thursday	12.1	14.0	9.4	11.5
Friday	17.0	6.5	23.9	16.2
Saturday	10.6	23.4	18.8	17.0
Sunday	9.9	7.5	12.8	10.1
Time of arrival				
Morning	23.5	41.6	24.1	29.3
12.01–3 p.m.	35.3	20.8	27.8	28.7
3.01–6 p.m.	23.5	24.8	32.4	26.7
6.01 p.m.–midnight	17.6	12.9	15.7	15.4
Previous visit				
Yes	68.3	76.2	75.4	72.9
No	31.7	23.8	24.6	27.1
Residence				
South Island	31.8	43	50.4	41.1
North Island	24.2	28	22.1	24.6
Overseas	43.9	29	27.4	34.3
Travel to Christchurch				
Car	56.8	62.1	68.6	62.2
Plane	32.4	22.3	26.3	27.5
Other	10.8	15.5	5.1	10.3
Travel within Christchurch				
Car	65.6	83.3	85.2	77.1
Walking	13.6	5.6	1.9	7.6
Other	20.8	11.1	13.0	15.5

(*Source*: Pearce, 1987d).

Table 4.6   Aspects of motel choice in Christchurch (November, 1985)

Variable	Central city (%)	Major routes (%)	Suburbs (%)	All (%)
Main reason for staying in motel				
Convenience	34.1	34.9	30.3	33.1
Habit	18.2	20.8	12.8	17.3
Facilities	12.9	7.5	5.5	8.9
A place to stay	7.6	7.5	5.5	6.9
Location	0.0	7.5	6.4	4.3
Other	27.3	21.7	39.4	29.4
Main reason for choosing particular motel				
Centrality	39.7	20.0	2.1	22.2
Proximity	2.2	14.0	29.7	14.4
Location	0	21.0	11.7	9.8
Facilities	21.3	16.0	15.3	17.9
Previous experience	6.6	9.0	6.3	7.2
Other	30.1	20.0	34.2	28.6
Means of selecting motel				
Previous experience	23.5	34.0	25.4	27.2
AA book	15.3	15.1	14.9	15.1
Yellow pages and other directories	16.0	3.8	10.5	10.7
Driving past	10.4	19.8	9.6	12.9
Referral	15.3	12.3	14.9	14.3
Other	19.4	15.1	24.6	19.8
Pre-booking				
Yes	59.3	63.6	60.2	60.8
No	40.7	36.4	39.8	39.2
Booking agent				
Guest	38.2	39.7	41.7	39.7
Friends/relations	5.6	13.7	19.4	12.4
Other	56.2	46.6	38.9	47.9

(*Source*: Pearce, 1987d).

the real holiday decision-making process takes place'. Advertisers and promoters of destinations and products, for example, can have little impact on critical demand factors such as free time and available income although they might benefit from an analysis of these factors in designing their marketing and development strategies. Hodgson also suggests second-order decisions concerning the type and purpose of the holiday are also predetermined to a large extent. Basic motivations would appear to play a significant role at this level. As was noted earlier,

Table 4.7 A hierarchy of holiday decisions

First-order questions
    Will a holiday be taken at all this year?
    When will the holiday be taken?
    How long will the holiday be?
    How much will be spent?
    With whom will the holiday be taken?

Second-order questions
    What type of holiday?
    What is the purpose of the holiday?

Third-order questions
    Where to go?
    What type of accommodation to use?
    What type of catering arrangements?
    Which operators to use?
    How to book?

Fourth-order question?
    When to book?

(*Source*: after Hodgson, 1983).

the basic need to get away or even fulfil a range of other more specific needs can usually be met at a multitude of destinations. With the basic style of the holiday set, the specifics are then established in terms of a host of third-order questions such as where the holiday will actually be taken and what form of accommodation will be used. Table 4.6 illustrates some of the details of this process. Hodgson (p. 22) argues that questions of this order are 'potentially eminently influenceable' and suggests:

> National tourist advertising can increase consumer interest in Greece
> at the expense of Portugal among those seeking a non-Spanish summer
> holiday; resort promotion may attract the regular visitor to Brighton to
> consider Bournemouth; the inclusive tour buyer may be influenced by the
> photographs and descriptions of Benidorm and may decide against Lloret
> de Mar.

A slightly different perspective is presented by van Raaij and Francken (1984) who draw on consumer decision literature in proposing a five-stage 'vacation sequence' consisting of: the generic decision, information acquisition, joint decision-making, vacation activities, and satisfaction and complaints. The generic decision is comparable to Hodgson's first-order questions and related determinants and involves deciding whether to take a holiday or to stay at home. However, van Raaij and Francken also observe that the decision to go on holiday or not may be closely linked to the type of trip taken, with some groups taking a less expensive holiday rather than foregoing their vacation entirely. Such modification of behaviour was noted earlier with respect to Belgian and West German vacations in the early 1980s (Boerjan and Vanhove, 1984). Meunier (1985) provides some interesting perspectives on this generic decision amongst low-income Quebec families, noting a fundamental difference between the employed for whom the decision to take a holiday is part of the work–vacation cycle, and those on social security, for whom it is not.

The second stage in the sequence is information acquisition, with information being sought about different aspects of the vacation: destinations, accommodation, transport and so on. According to Jenkins (1978), holidaymakers select the destination first then decide on the type of accommodation. Different sources of information may be used for different types of holiday and different facets of the trip (e.g. Table 4.6) but personal knowledge and the influence of family and friends is often shown to be more significant than formal promotional material or the advice of travel agents (Walter and Tong, 1977; Gitelson and Crompton, 1983). In any event, potential tourists will not usually have perfect knowledge nor be fully aware of all the destinations (or elements within those destinations) which might be able to fulfil their motivations or satisfy their needs. Van Raaij and Francken cite the work of Woodside and Sherrell (1977) who distinguish several sets of destinations. 'Evoked', 'inert' and 'inept' destinations are all places which people are aware of but which they are, respectively, likely to visit, be undecided about visiting or unlikely to visit because of some negative evaluation. The average evoked set for a small sample of tourists in South Carolina was found to consist of just under four destinations. Woodside and Sherrell also distinguish destinations which are 'unavailable' through cost, access or some other restriction from those people are 'unaware' of and suggest different marketing and development strategies will be needed to remedy each of these. Other researchers have attempted to assess the images and perceptions which people have of particular destinations so that appropriate marketing activity might be undertaken to improve potential tourists' knowledge and to bring their image in line with what the destinations actually have to offer (Hunt, 1975; Goodrich, 1977; Haahti, 1986).

Citing the work of Davis and Rigaux (1974) and Jenkins (1978), van Raaij and Francken conclude that vacation planning and decision-making are essentially undertaken jointly by spouses, with some input from children. This was especially the case with choice of the vacation destination, a finding supported by other studies (Myers and Moncrief, 1978; Ritchie and Filiatraut, 1980). However, other aspects of the decision-making process may be male dominated, notably the amount to be spent, the timing of the holiday and the selection of the route.

The actual process of making decisions will vary from tourist to tourist. 'For some consumers, the decision-making process is a long sequence of information acquisition and comparison of alternatives. For others, it may be an impulsive "last minute" decision. For a self-organized vacation, it is in fact a sequence of subdecisions about meals, excursions and other activities' (van Raaij, 1986, p. 4). For other forms of tourism different decision-making processes will be relevant. Fortin and Ritchie (1977) provide a useful study of convention site selection which highlights the role of elected officers in decision-making and shows that different types of associations give weight to different factors in choosing the location of their conference.

Aspects of tourist activities and satisfaction, the fourth and fifth stages in the vacation sequence, have been reviewed in earlier sections where it was shown analyses of tourist benefits and time–budget studies, amongst other approaches, provided useful indirect measures of motivation. Van Raaij (1986) points out that satisfaction (and dissatisfaction) have an important bearing on future vacations, not only for the individuals concerned but also for others given the influence of friends and relations' advice and views in decision-making.

The perspectives of Hodgson, van Raaij and Francken, Jenkins and others provide a useful overview of tourists' decision-making processes. As in other instances, more detailed information relevant to specific destinations or products will be needed to supplement specific marketing, management or development decisions. The data in Table 4.6, for example, suggest moteliers in central Christchurch seeking to extend their overseas visitor clientele should place greater emphasis in their advertising on the facilities they have to offer and develop or strengthen their contacts with travel agents through whom many reservations are made. Others, particularly those on major routes, might want to reappraise the value they get from advertising in directories other than the Automobile Association handbook.

MARKET SEGMENTATION

Analysis of global or large-scale patterns of demand, such as those depicted in Figs. 4.3 and 4.4, can be very useful for establishing general trends and assessing overall market conditions. Such analysis will be important, for example, for national tourist organizations or large multinational companies. Most developers, however, will only be catering for a small share of any given market. In view of the diversity of tourist demand already noted, it will therefore be important in most cases to relate the needs and preferences of particular segments of the market to specific destinations or products.

Davis and Sternquist (1987) note that for market segmentation to be effective the segments must be:

1. Measurable – a distinct group must be identified.
2. Assessable – promotional efforts must elicit the desired response.
3. Substantial – the segment must be large enough to warrant the expenditure of time, money and effort to attract it.
4. Reliable – consumer characteristics must be stable indicators of market potential.

Given the different ways in which tourists may be characterized and the composite nature of the tourism experience, there are many possible ways of segmenting the vacation market. Oppendijk van Veen and Verhallen (1986) identify three basic approaches to market segmentation: analysis of consumer characteristics, analysis of consumer response and the simultaneous analysis of both these sets of factors.

In the first approach consumers are grouped according to some set of attributes, with the different groups being subsequently related to behavioural differences (i.e. forward segmentation). Initially emphasis was given to standard socio-economic variables such as those depicted in Table 4.4 but as this has not always discriminated effectively between groups, growing attention has been given to psychographic segmentation which takes into account less tangible factors such as attitudes, interests, opinions and lifestyles (Hawes, 1977; Perreault, Darden and Darden, 1977; Shih, 1986). A survey of potential New Zealand travellers to Australia, for example, identified in this way four major segments: the 'older relaxation/comfort seekers', 'young excitement seekers', the 'nature/culture' group, and the 'unadventurous sightseer' (Burfitt, 1983). Different marketing strategies are then designed for each of these groups whose travel requirements differ from one another.

With the second approach, consumers are assigned to groups 'on the basis of their similarity in behavioural response to the supply of goods and services (e.g. the chosen type of vacation)'. The differences between the groups are then related to consumer characteristics (backward segmentation) as in Tables 4.5 and 4.8. Segmentation of consumer response has been undertaken in a variety of ways including: destination activities (Graham and Wall, 1978), benefits (Mazanec, 1984), frequency of visit (Kaynak, Odabasi and Kavas, 1986), distance travelled (Etzel and Woodside, 1982), travel patterns (Pearce and Johnston, 1986), seasonality (Calantone and Johar, 1984), types of tour packages (Thomson and Pearce, 1980) and time-share ownership (Woodside *et al.*, 1986). Taylor (1986) reports on a multi-dimensional segmentation of the Canadian pleasure travel market involving travel philosophies, benefits and activities and interests.

The third approach identified by Oppendijk van Veen and Verhallen – the simultaneous analysis of consumer characteristics and response – is much less common. Assigning consumers to groups on the basis of the relationships between their characteristics and their behavioural response to the supply of goods and services is much more complex analytically. In their survey of Dutch holidaymakers, Oppendijk van Veen and Verhallen employed canonical analysis to establish these relationships directly and subsequently identified seven overlapping segments: the organized vacation, the beach vacation, the domestic vacation, vacation with children, the one-to-two-person vacation, the long camping vacation and the short vacation.

From a broader development perspective one of the most useful types of market segmentation is that which enables specific types of impacts to be related to particular groups of tourists and different types of consumer response. In this way demand can not only be related to particular tourist products but to the consequences of developing those products and segmentation of the market can suggest different strategies depending on the development goals sought (Pearce and Johnston, 1986).

Pearce and Johnston segmented travellers to Tonga on the basis of their travel patterns within the kingdom, that is according to whether they restricted their stay to the main island of Tongatapu or combined a stay there with a visit to one or more of the other major islands or groups, notably Vava'u, 'Eua and Ha'apai. Table 4.8 shows that the two groups of visitors, Tongatapu only (TO) and Tongatapu plus (TP) have distinct profiles. Some of the differences between the two groups can be attributed to the varying purposes of visit but it seems those going beyond Tongatapu constitute a group of more adventuresome, independent and informal travellers whereas those limiting their stay to the main island are much more conventional tourists who rely more heavily on the tourist industry infrastructure.

Table 4.9 indicates clearly that the impact of the TP visitors is much greater than their share of the market initially suggests. Although they comprise only a fifth of the respondents, the TP group accounted for half of all the bednights generated by the sample in the kingdom, with their mean length of stay being four times greater than that of the TO visitors. These differences in length of stay also contribute to variations in spending between the two groups. The average per person per night expenditure is higher amongst the shorter stay TO group, but their mean total expenditure is less than half that of the TP visitors. Consequently the share of total expenditure generated by the TP visitors is again significantly greater than their share of the market (44.6 per cent cf 21 per cent). Moreover, the money spent by the TP group would be more widely dispersed both geographically and amongst a greater range of businesses, accommodation included.

Table 4.8 Profiles of visitors to Tonga by islands visited (1981)

Profile item	Tongatapu only (n = 170)	Tongatapu plus (n = 40)
	%	%
Country of residence		
New Zealand	44.7	30.8
Australia	11.9	35.9
Other Pacific	20.2	12.8
USA/Canada	9.5	2.6
Others	13.7	17.9
Purpose of visit		
Pleasure	55.7	68.6
Visit friends or relatives	3.8	11.4
Business or work	40.5	20.0
Type of travel		
Tour	40.8	0
Independent	59.2	100
Number of visit		
First	72.8	68.6
Second	8.1	25.7
Third or more	19.1	5.7
Main type of information used		
Travel agent	45.0	25.7
Guide book	10.6	11.4
Friends and relatives	19.4	45.7
Other	25.0	17.1
Accommodation Tongatapu		
Hotels	83.4	44.4
Other accommodation	16.6	55.6

(*Source*: Pearce and Johnston, 1986).

If further expansion of tourism is seen as a desirable way of developing the kingdom's economy, then one of the major choices facing policy-makers is the extent to which future tourist activities continue to be focused on Tongatapu or dispersed more widely throughout the group. If the latter goal is pursued then different strategies will be needed for the two market segments identified earlier. The TO group are influenced more by travel agents but at present travel agents' familiarization visits are limited solely to Tongatapu and accordingly they are unlikely to promote the outer islands to their clients. Extending the agents' visits may in turn expand their clients' itineraries. At the same time, the majority of the TO group at present appear to express

Table 4.9 Impact of visitors to Tonga by islands visited (1981)

	Tongatapu only (*n* = 146)	Tongatapu plus (*n* = 39)
Respondents	79%	21%
Total nights in kingdom	49%	51%
Mean nights in kingdom	5.3	20.7
Total nights on Tongatapu	66.4%	33.6%
Mean nights on Tongatapu	5.3	10
Total expenditure in kingdom	55.4%	44.6%
Mean expenditure per person in kingdom	US$205.26	$484.51
Mean expenditure per person per night in kingdom	US$49.10	$30.38

(*Source*: Pearce and Johnston, 1986).

a preference for good quality hotel accommodation and encouraging them to visit other islands may also involve the provision of more and better accommodation there. Conversely, differences between Tongatapu and other parts of the kingdom might be consciously promoted to the two distinct segments by encouraging the development of smaller less formal facilities on selected outer islands. Given its potential economic impact the TP group (the multiple-island visitors) certainly merits further attention. As noted in Ch. 3, experiments with alternative tourism elsewhere in the South Pacific have frequently met with limited success, but in these cases there is little evidence of any marketing preceding such developments.

FORECASTING DEMAND

Examples throughout this chapter have illustrated the interrelationships between demand and development and shown how knowledge of the former can be used effectively in decisions regarding the latter. Where the concern is not just with explaining past patterns or analysing existing situations but with planning the path future tourist development might take, then an awareness of future patterns of demand becomes very important. Decisions regarding investment in basic tourist infrastructure or development of certain sorts of facilities will require estimates of likely demand at a given period, for example 1, 5 or 10 years in the future.

Given the variety of measures of tourist demand and types of tourism to be found it is not surprising that a wide range of methods of forecasting the demand for tourism have been employed. Several very useful and comprehensive reviews of these methods have recently been undertaken (Archer, 1980; van Doorn, 1986; Esteban Talaya and Figuerola Paloma, 1984; Uysal and Crompton, 1985; WTO, 1981) and only the main considerations are summarized here.

Demand forecasting techniques can be divided into two basic types, qualitative and quantitative. Qualitative methods 'depend upon the accumulated experience of individual experts or groups of people assembled together to predict the likely outcome of events' (Archer, 1980, pp. 9–10) while quantitative ones involve the numerical analysis of relevant data sets.

Qualitative forecasts range from those involving the best guesses of a group of tourist industry professionals or others sitting around a table to those based on more structured approaches such as the Delphi technique (Kaynak and Macaulay, 1984) and scenario writing (Choy, 1984). Qualitative approaches, especially the more structured ones, can be particularly useful for analysing medium- and longer-term changes in demand, particularly where the concern is not only with estimating absolute levels of demand but with identifying changes in the composition of that demand by including less tangible factors such as motivation and taking account of changes in causal factors over time. Some of the reports by the WTO (1981 and 1985) provide useful examples of these general changes. Qualitative approaches are also usually less demanding in terms of data, time and costs.

Quantitative forecasting, on the other hand, requires the existence of an adequate data base to which a range of numerical techniques can be applied. Archer (1980, p. 5) categorizes these techniques into two basic types: 'time series, i.e. those which involve the statistical analysis of past data of the variable to be forecast; and causal, i.e. based on the statistical analysis of data for other variables shown to be related to the one of interest.' In the first, for example, tourist arrivals at some future date are estimated by analysing past arrivals and projecting these trends into the future with varying degrees of sophistication. In the second, changes in tourist arrivals are calculated on the basis of changes in causal factors or the determinants of demand using econometric modelling and techniques such as multiple regression. In these models, the forecasts depend on the relationships between the factors remaining constant and for this reason are most applicable to short-term forecasts up to 2 years though longer forecasts have been made, sometimes with variations in key assumptions being taken into account (Edwards, 1985).

Quantitative techniques can be particularly useful for forecasting demand at existing destinations but are obviously limited in their application to new resorts or new projects (e.g. theme parks, tourist towers or gondolas) where there are no existing data. In these cases, recourse has to be made to analysis of more general patterns of demand, consideration of levels of demand for similar projects elsewhere or surveys of potential visitors to ascertain their interest and willingness to pay. The question of scale is also important, for changes in demand at the national level (where data on international tourist arrivals are most readily available) may not be reflected uniformly throughout the country in question. Different national markets or segments have different preferences so that those regions or resorts appealing to the growing markets or segments will experience a different pattern of growth to those favoured by markets which are stable or in decline (Pearce, 1987a).

A study of wide applicability and interest is that by Edwards (1985) which contains a set of forecasts of international tourism to 1995 based on a series of separate econometric models for the twenty main origin countries in 1983. The main factors taken into account are projected rises in real household disposable income, projected changes in the cost of travel abroad relative to domestic prices (including exchange rate movements), a 'constant' or time-related growth factor not attributable to these first two factors and a 'ceiling' to travel abroad.

Edwards predicts an increase in real forecast expenditure including fares of 7 per cent per annum for the period 1983–90 (cf. under 6 per cent per annum for 1975–83) increasing to 8 per cent a year during 1990–95. World tourism expenditure (including fares) is predicted to increase from $154 billion in 1983 to $367 billion in 1995; the number of international trips made is expected to rise from 535 million in 1983 to 784 million in 1995 and the number of nights spent abroad is forecast to pass from 3083 million to 5014 million over the same period. The differential rates of increase on these different measures is due in part to a shift to more long-haul travel and an increase in the relative importance of certain high-spending markets such as Saudi Arabia. By 1995 Japan is expected to become the world's second most important market (after the FR Germany) in terms of expenditure on international tourism, pushing the USA and UK down to third and fourth positions, respectively. Germany is also predicted to retain the number one position in terms of nights spent abroad and replace the USA as far as the number of trips abroad is concerned. Significant changes in the composition of the other 'top ten' countries on these measures are also forecast.

CONCLUSIONS

Research on the development of tourist destinations needs to be complemented by interrelated studies of demand, many of which will be focused on conditions in the markets or origins. Whether this concern is with understanding past processes of development or planning for the future expansion of tourism (Ch. 7), more attention needs to be directed at the nature of tourist demand. This chapter has emphasized the complex nature of the phenomenon, highlighting the role of less tangible factors such as those influencing tourist motivations and other more concrete ones such as incomes and paid holidays in creating a far from homogeneous market. While certain basic factors have been identified – the desire to get away, the need to have the time and money to do so – many combinations of these and more specific variations on them exist to produce a demand for many different types of tourist travel and tourist product. Moreover the demand is not constant but ever evolving, generally increasing in volume while undergoing qualitative changes.

Despite the complexity of tourist demand, important steps have been made in its analysis, with various measures of demand being established, useful advances being made in the understanding of motivations and increasing information becoming available relating to peoples' ability to travel. More applied work has also been undertaken relating to decision-making, market segmentation and demand forecasting. Examples have been given throughout of the development implications of various aspects of demand discussed (e.g. Crompton, 1979; Boerjan and Vanhove, 1984; Pearce and Johnston, 1986) but in all these areas the links between demand and development need to be explored more thoroughly and made more explicit. Origin and destination, demand and supply, are not separate and isolated phenomena but integral parts of the same tourist system (Fig. 1.1).

Evaluating tourist resources

Analysis of demand in the markets must be complemented by an evaluation of tourist resources in the destination. Given the multi-faceted nature of tourism and the complex nature of tourist demand, many different resources might be exploited for tourism and many different factors may influence where tourist development occurs. As tourist resources do not occur evenly or randomly in space, developers and planners will be faced with such practical questions as assessing the feasibility of developing a particular site, the selection of one site from a number of alternatives for a specific project or the broader evaluation of an area in terms of general tourist potential. The place of such evaluation in the overall planning process is examined more fully in Ch. 7. The present chapter discusses first of all the various locational factors which influence tourist development and how these may be assessed. These individual factors are then brought together in a general review of techniques and methodologies for evaluating tourist resources. The chapter concludes with a short case study which examines the relative importance and interaction of the locational factors related to ski-field development.

LOCATIONAL FACTORS

The factors influencing the location of tourist projects or the tourist potential of an area can be grouped into seven broad categories – climate, physical conditions, attractions, access, existing facilities, land tenure and use and other considerations such as the availability of regional development incentives. These factors are interrelated and the categories are not wholly exclusive. Climate, for example, may be an attraction; the attractiveness of an area may depend in part on its access and certain forms of land tenure may be subject to various constraints. Nevertheless, the categories outlined do provide an appropriate focus for the literature reviewed as well as a useful introduction to the methodologies discussed subsequently. The importance of any one of these factors will depend on the type of development and especially on the scale of analysis. Climatic data, for example, may suggest broad regions which may be suitable for developing tourism, but within a resort in any one of these accessibility will be a prime determinant for the successful location of a specific hotel or motel. The importance of these factors will also vary from group to group. The price of land and nature of land tenure will be crucial to developers though not to most tourists, but it will be the latter group's appreciation of certain attractions which will influence where the developer wishes to acquire land. Thus while resource evaluation is closely linked to demand and marketing studies, other development considerations also come into play. The selection of an area or the scale of development proposed will also depend on the extent to which the area may be developed. Studies of carrying capacity may thus form an integral part of the evaluation process and be of interest not only to the developer but of concern to the host community who may have to bear the externalities when various thresholds are exceeded. Such studies are reviewed at the conclusion of this first section.

Climate

Although the relationships between tourism and climate have long been recognized, comparatively little research has been directed specifically at the nature of these relationships (Mings, 1978a). Climatic variations affect tourist development in a number of ways, outlined below.

Attraction

The exchange of a home climate for a more favourable one elsewhere is one of the more prominent forms of inversion noted in Ch. 4 and one which underlies the popularity of sunlust tourism to Mediterranean, Caribbean and Pacific destinations. It is simply pleasant and agreeable to spend one's holidays in an area characterized by warm temperatures and high sunshine hours. This 'hunt for the sun', however, is a comparatively recent phenomenon. Although the mildness of the winter climate of the Riviera has been appreciated since the eighteenth century, it was considered unhealthy to remain on the coast into July and August and it is not until the mid-1930s that the summer visitors surpassed the '*hivernants*'. This example serves well to emphasize at the outset that resources are, after all, cultural appraisals. Other resorts owe, or owed, their popularity to mild or invigorating climates, such as the hill stations of India and the Far East (Robinson, 1972; Senftleben, 1973) and Europe's *stations climatiques* like Leysin and Superbagnères. More specific forms of tourism, such as that based on winter sports, will depend heavily on other climatic criteria. For other forms still, such as sightseeing or cultural tourism, a favourable climate may be an important secondary factor. Conversely, adverse climatic conditions will detract from an area's interest or beauty and the absence of certain conditions, for example snow, will prohibit certain activities altogether.

Seasonality

Favourable climatic conditions for any specific activity will often occur only during certain seasons. The degree of seasonality will have a marked bearing on an area's profitability. Longer seasons give a better utilization of plant and equipment and consequently yield higher returns on capital invested. Blessed are those regions with a 'second season'. Seasonality will become more important as the dependence on climatic factors increases. The length of season will be more critical for coastal and winter sports resorts than for capital city tourism. In India, Singh (1975, p. 42) notes that: 'the odyssean spirit of wanderlust seems to be overcoming the climatic barriers'. There is little variation in tourist arrivals from one season to another and religion-based tourism appears wholly independent of climatic comforts. Singh also notes how the richer tourists can afford air-conditioning and other amenities reducing the influence of adverse climatic conditions.

Construction

Development costs will rise where the construction period is limited by seasonal climatic constraints such as rainy seasons or harsh winters. Additional costs will also be incurred where extreme temperatures increase the need for central heating or air conditioning (Mings, 1978a). In the case of La Plagne (Fig. 3.3), the physical integration of the resort is in part a response to the practicalities of heating buildings located at 2000 m and the functional form of the apartments is a reflection of the shortened construction season. Microclimatic considerations are also important in the design of coastal resorts; for example, siting swimming pools so that they are not constantly in the shadow of high-rise hotels.

Operations

Short-period climatic events may hinder the operation of certain facilities. High winds may lead to the closure of aerial cableways and severe storms will limit yachting or sightseeing cruises and may disrupt access. Strong coastal breezes may give rise to unpleasant conditions on sandy beaches and recurrent ground fogs will restrict aircraft operations. The length of activity-day may be severely limited in harsh winter regimes.

The geographical problem here is to identify those areas or regions within a given study area which are climatically suitable for some form of tourist development or recreational activity. Several attempts have been made to derive recreational climatic classifications through grouping a number of elements deemed to affect recreational activity and tourist comfort.

In his classification of the Canadian North-West Territories, Crowe (1975) selected three factors for each of the two main seasons (*winter* – length of activity-day, temperature, wind; *summer* – temperature, cloud cover, wind) and defined for each limiting values for 'ideal', 'marginal' and 'sub-marginal' classes. These factor classes were then combined to give four classes of outdoor activity potential, with each factor given equal weighting. A similar approach was adopted by Day *et al.* (1977), to map variations in suitability for winter sports and summer activities in Bay of Fundy National Park. Four parameters were used for classifying areas of climatic suitability for skiing and snowmobiling – temperature, wind speed, precipitation and visibility. Areas described as most suitable meet all four criteria, those described as climatically least suitable meet one or none (Fig. 5.1).

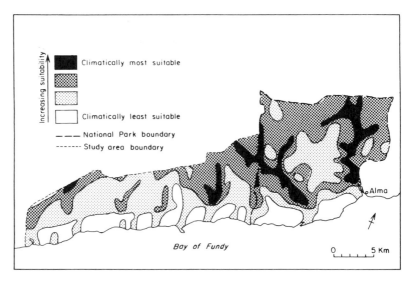

Fig. 5.1 Climatic zones related to snowmobiling and skiing, January, Bay of Fundy.
(*Source:* Day *et al.*, 1977).

Where more adequate data are available, a more quantitative approach can be adopted. In his recreational climatic study of Ontario, Crowe (1975) considered two main factors – comfort and weather – for a range of summer activities, e.g. landscape touring, vigorous activity and 'beaching'. Comfort was expressed in terms of maximum temperature and humidity and weather in terms of cloud cover and precipitation. Percentiles were then calculated for each factor and combined into a general index for each activity. The results of the summer analysis indicated that conditions for beaching were much better in Southern Ontario and that other parts of the province were more favourable for landscape touring and vigorous activities.

A different approach was used by Harker (1973) in her classification of the climatic potential of the whole of Canada for winter recreation. Seven factors were selected and the values for each given different weightings (e.g. mean annual total snowfall was considered more important than humidity). The weighted scores for each of the seven factors were then summed for fifty points throughout Canada, and mapped to give an isopleth map showing the climatic suitability of different parts of the country for establishing winter sports (Fig. 5.2).

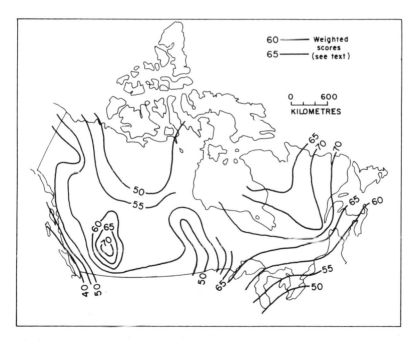

Fig. 5.2 Climatic potential for winter recreation in Canada. (*Source:* Harker, 1973).

A general correspondence was found between the areas identified as having a high potential and those where ski-resorts were already well established.

Besancenot, Mounier and de Lavenne (1978) use a similar technique in their attempt to classify summer climates suitable for tourist development. However, instead of summing the values of the six parameters used, a range of values is defined for each parameter to give a ninefold classification. The parameters used – sunshine hours, cloud cover, maximum temperature, wind speed, vapour pressure and precipitation – are said to combine a measure of comfort with one of attractivity. These writers also reject the use of average values, such as monthly means, in favour of actual daily recorded values, as they argue it is these which are the most meaningful for the tourist. This information is then presented in histograms for 10-day periods for a number of Breton and Mediterranean resorts. However, when it comes to mapping summer conditions at a regional level this is done simply in terms of the number of fine days per month. Besancenot (1985) subsequently applied the same approach in a comprehensive classification of the coasts of the Iberian Peninsula.

Besancenot *et al.* also stress that tourists are concerned with how reliable or variable resort climates are. Climatic variability was measured by the use of the coefficient of variation, values for which were shown to decline over the summer months with latitude. Moreover, variability diminishes where climatic conditions are the most favourable. However, Dauphiné and Ghilardi's (1978) review of climatic comfort indices with reference to the Côte d'Azur shows that the extent of seasonal variability depends in part on the index used. Clausse's index shows little variability in summer but much in winter whereas the converse is true of Terjung's.

A further recreational comfort index is proposed by Yapp and McDonald (1978), using a model of the heat balance of the human body and based on the maximum use of actual observations and minimum use of average values. As the human heat balance depends in part on the activity level, the model was applied to different recreational activities such as sunbathing, walking and boat fishing and the frequency of conditions that are pleasant, indifferent or unsuitable for these was calculated. Mieczkowski (1985) describes an attempt to derive a global tourism climatic index based largely on comfort considerations.

Although as yet far from perfect, these attempts to define tourist climates suggest that it is insufficient merely to consider a single element, such as temperature, or a series of elements, as has often been the case in the past. Some attempt must be made to combine particular parameters into a more general index and one which is related specifically to particular recreational or tourist activities. The measures discussed here show it is possible to identify spatial and temporal variations in such indices. In this last respect it is important to note that periods identified by several writers as being climatically suitable for tourism were significantly longer than the period of actual tourist use, a reflection in part of the constraints of institutionalized holidaytaking discussed in Ch. 4.

Other recent work in Canada has focused not so much on spatial variations in climate but on temporal changes, notably the implications of CO_2-induced warming aspects of tourism and recreation in Ontario (Wall *et al.*, 1986a, 1986b; McBoyle and Wall, 1987). Different scenarios suggest that as a result of rising carbon dioxide levels, Canada may experience a generally warmer, drier climate by the second half of the twenty-first century. The implications of these predicted changes vary according to the type of tourism considered. The length of the camping season in Ontario would be significantly extended, while ski-resorts in the Laurentians would experience a serious reduction in reliable snowcover. In the latter case, a minimum financial loss of \$10 million is estimated with the ski industry in the area being virtually eliminated under conditions of maximum change. The authors note that the scenarios involved are best regarded as possible futures rather than predictions of the future but suggest that ski-resort operators should take account of climate change and that 'a program leading to diversification could be a wise investment for the future' (McBoyle and Wall, 1987, p. 50).

Physical conditions

Physical conditions, other than climatic ones, are important in several aspects of tourist development, particularly at the resort level.

Building site

Firstly, there must be a sufficiently large site to locate the required accommodation, supporting facilities and necessary infrastructure. Soils, geology, topography, slope stability and aspect will be among the many physical site attributes to be considered. Additional factors will include the possibilities for water supply, drainage and sewage disposal.

Access

Ease of access, as determined by these attributes, will also be important, particularly for winter sports resorts and development based on natural attractions. Modern technology can overcome access to most sites as indeed it can enable construction in most places. The costs involved, however, may be prohibitive.

Recreational resources

The recreational base of many resorts depends on their physical attributes. Georgulas (1970), for instance, proposes the following criteria for a first-class beach:

1. *For passive activities.* A combination of the following: (a) Beaches – fine clean sand, 300 ft minimum length and 50 ft minimum width. Beach should be relatively free of exposure for at least 80 per cent of the year. (b) Backland – shade, tree, hospitable environment, free from man-made nuisances (e.g. dumping) and natural nuisances (e.g. poisonous insects, snakes). Slope less than 15° for easy access and potential for development.

2. *For active activities.* Water quality: No (or very little) silting; colour, less than 5 Hazen units, odourless, coliform count less than 50 per 100 ml, free of biological nuisances, sea-bed free of coral and sharp rocks to 8 ft depth during high tide, free of dangerous currents. The beach and immediate water area should have a gradient of not more than 8°. Beach quality as in (1) but longer and wider. Swimming should be possible for 9 months of year.

Ski-field development will of course depend on other physical criteria (see below) as will the construction of marinas. More generally, other physical features are major attractions in themselves; for example, the Grand Canyon in Arizona, Franz Josef Glacier in New Zealand and the Iguazu Falls along the Brazil/Argentine border. Classification of attractions constitutes a problem in itself.

Attractions

'Tourist development is a problem of matching naturally or historically given resources to the demands and preferences of actual or potential tourists' (Piperoglou, 1967, p. 169). However, considerable variety exists in tourists' demands and preferences as was shown in Ch. 4 and there is no universal measure of tourist attractiveness. For the purposes of resource evaluation and spatial differentiation, the task facing the geographer, planner or developer is no less than 'to reduce phenomena of aesthetic or cultural significance to quantifiable magnitudes for purposes of comparative evaluation' (Piperoglou, 1967, p. 169). Piperoglou outlines four basic steps in evaluating tourist attractions:

1. Survey the market to discover tourist preferences.
2. Identify and evaluate what tourists want in the area of study.
3. Define regions in terms of the spatial interplay of resources.
4. Check the capacity of the study region to absorb visitors in terms of both the human and space factor.

The first three of these will be dealt with here, the fourth being considered in the section on carrying capacity.

Thus for his study of western Greece, Piperoglou undertook a survey of visitors to Greece. Three main groups of attractions were identified: 'Ancient Greece', 'picturesque villages' and 'sun and sea'. Resources in the study area were then evaluated and plotted on a series of overlay maps, each resource being represented by a circle proportional to the preferences expressed in the survey. In evaluating particular resources, attention was paid to the extent to which it was unique or occurred elsewhere in Greece. Tourist regions were then defined on the basis of the coincidence of different resources within a given area, in this case within an 80 km radius of a base point. Such a distance was considered the average a tourist would willingly travel on a half-day trip to a point of interest. Where two of the three main resources were present, the circles were expanded to their square and to the third power when all were evident. A fourth weighted element, an infrastructural component in the form of an urban settlement of 50,000 people or more, was also introduced. Scores for each region were then summed and development priorities determined on the basis of the total score of each region. This approach then combines the weighting of various attractions with the spatial association of these. The clustering of resources is especially important, the whole being greater than the sum of the individual attractions, especially where a range of different resources occurs.

Visitor surveys can be costly in terms of both time and money. A second technique is to survey expert opinion, commonly using the Delphi method, as a surrogate for tourist preferences. In seeking to establish the importance of culture as a determinant of tourist attractiveness, Ritchie and Zins (1978) contacted some 200 'informed individuals'. Respondents were asked by questionnaire to consider a range of factors from the stand-point of an 'average traveller'. Responses were then measured both on ordinal rank scales and on eleven-point interval scales. In general terms, cultural and social characteristics were found to rank second behind natural beauty and climate but ahead of other factors such as accessibility and attitudes towards tourists. The relative importance of the different socio-cultural elements was then determined for residents and non-residents (Table 5.1). Unfortunately, in this article this first stage was not related to the actual evaluation of resources in a specific region.

In his comprehensive survey of tourist resources in Southern Africa, Ferrario (1979) combines features of each of these methodologies. Firstly, 2300 different features mentioned in ten guidebooks on South Africa were inventoried and classified into twenty-one categories. These were then evaluated in terms of two criteria – appeal and availability, using the formula $I = (A+B)/2$, where I is the index of tourist potential,

Table 5.1 Relative importance of socio-cultural elements influencing the cultural attractiveness of a tourist region.

Factor	Ordinal rank of importance		Average importance on interval scale		Interval order of importance		Overall order of importance	
	R	N-R	R	N-R	R	N-R	R	N-R
Leisure activities	1	1	9.18	8.54	1	2	1	1
Gastronomy	2	3	8.63	8.32	2	1	2	2
Handicrafts	3	2	7.95	8.66	4	3	3	3
Traditions	4	4	8.34	7.80	3	5	3	4
History	5	5	7.65	8.24	6	4	5	4
Art/music	6	6	7.66	7.38	5	6	5	6
Architecture	7	7	7.01	7.22	7	7	7	7
Work	8	8	6.19	6.88	8	8	8	8
Language	9	9	5.55	5.64	9	9	9	9
Education	10	10	4.77	4.87	10	10	10	10
Dress	11	11	3.73	4.56	12	11	11	11
Religion	12	12	4.02	3.87	11	12	11	12

(Source: Ritchie and Zins, 1978).
R = Residents
N-R = Non-residents

A is the appeal component, or demand, *B* is the availability component, or supply.

Demand was first assessed by a large scale visitor survey. A strong preference was shown amongst the twenty-one categories listed for environmental features, namely scenery and landscape, wildlife and natural vegetation. The percentage of preference received for each category was taken as an index of tourist demand and reduced to a scale of from 1 to 10 (e.g. 77 per cent became an index of 7.7). Each of the 2300 individual attractions in Southern Africa was then evaluated by weighting its category index by a guidebook coefficient, that is by how many of the ten guidebooks reported it. To bring back the resulting value to a 1 to 10 scale, the square root of the product was taken. With the inclusion of this new weighting coefficient (*G*), the formula thus became $I = \sqrt{(AG + B)/2}$.

The *B* values or the index of accessibility were determined by the use of six criteria said to affect supply: seasonality, accessibility, admission, importance, fragility and popularity. 'Community influentials' throughout the country were then asked to rank individual attractions in their area in terms of these six criteria according to a descriptive nominal scale. These responses were subsequently transformed into a weighted numerical index and the different evaluations received by each attraction were then averaged. After plotting these values, clusters of attractions were identified for grid cells and further weighted by attendance figures on the principle that the sum of many low indices in a grid cell representing a cluster of less important features, could not be numerically equivalent to the presence in another cell of a single high index of a leading attraction. Finally, twenty tourist regions were identified in the combined territory of South Africa, Lesotho and Swaziland by the clustering of grid cells. These were found to correspond well with existing patterns of demand, though this is perhaps not surprising given the final heavy weighting by attendance figures.

Other writers have drawn attention to the variations in how different groups perceive and rank attractions. In a study of Chinatown, Singapore, Tieh (1988) found tourists emphasized the appeal of the 'people' element while local residents did not see themselves as attractions and stressed the interest of buildings, street scenes and shops. Witter (1985) compared the evaluation of Traverse City's, North Michigan, attractions by local retailers to those of visitors to the resort and showed the former was much more favourable. Moreover, the retailers attached greater importance to most attributes examined than did tourists in selecting a resort area. Witter concluded (p. 19):

> Differences between the tourist and retail cluster profiles indicate that either the target market is not well defined in the mind of the retailers or the sample of tourists surveyed is not representative of the area's actual tourist

clientele. An increased understanding of the multi-faceted needs of tourist customers is needed before residents can effectively market their resort area.

Although not without their limitations, these papers do suggest that is is possible to evaluate reasonably objectively aspects of attractiveness and to delimit spatial variations in their occurrence. In general, however, these studies have focused on specific identifiable attractions and with the exception of Witter (1985), there has been little attempt to incorporate the less tangible qualities of an area that might enable tourists to escape and to rest and relax.

Access

Two associated types of access are important in assessing potential locations for tourist development: physical access and market access.

Physical access will depend to a large degree on the existing infrastructure – the location of access routes, highways and railways, the proximity of airports. As such infrastructure is costly to provide, its presence or absence is very significant. The extent to which timetables and route schedules service destinations already equipped is also critical. As the development of aircraft technology has enabled the growth of longer and longer direct flights, some localities, especially island destinations, are now being 'overflown' in the same way that many small localities previously served by rail are now bypassed. Relative isolation will only be an advantage where the emphasis is on a luxury market and the exclusion of mass tourism.

Accessibility is also a feature of proximity to the market, whether measured in terms of travel time, cost or distance. At the international level, proximity to industrialized and urbanized countries with high standards of living will be important. The nearness of large urban areas and the characteristics of their populations (age, income) will be deciding factors at the national and regional level (Ch. 4). Different types of development will demand differing degrees of market access. Many second homes are used not only during long vacations but also at weekends. Proximity to the primary residence is therefore important and the majority of second homes appear to be within 100–150 miles (160–240 km) of major population centres. However, distances vary from country to country, being greater in the United States and smaller in Sweden (Coppock, 1977). For more specialized attractions, such as ski-fields and outstanding natural features, tourists will be prepared to travel greater distances. For developments which do not constitute a tourist ensemble in themselves, proximity to the market becomes particularly critical. Such is the case with the location of pleasure ports, the demand for berths depending on the resident population and the number of second homes in a region. In the Var (France), the ratio is approximately one berth for boats greater than 2 tons for every ten second homes (Perret and Bruère, 1970).

Much work on predicting flows to particular resorts and developments has centred on the use of gravity models (Archer and Shea, 1973; Smith, 1983). The basic gravity model expressed the flow of people between two centres as a function of the population of each and the distance between them. Tourist flows, however, are not necessarily reciprocal as Wolfe (1970) points out in the case of second homes where the traffic outward from an urban area to a cottage resort is not complemented by a reverse flow to the city. Consequently, when applied to tourist flows, the basic gravity model has been modified in various ways. Such modifications include the incorporation of variables measuring the attractiveness of the destinations (the number of ski-lifts, the ratio of water to land area) or the measurement of distance in terms of time or travel cost. Modified gravity models have been applied with some success to specific cases such as the demand for second homes and ski-fields (Bell, 1977; McAllister and Klett, 1976) and to tourist travel to Las Vegas (Malamud, 1973).

In many cases, it will be not so much actual distances which will determine an area's potential market as its location relative to other resorts or attractions. Several major projects have been developed as 'intervening opportunities', exploiting their location between existing resorts and the market. Developers of the Languedoc–Roussillon littoral, for example, sought in part to intercept holidaymakers heading for Spain. Cancun in Mexico has been promoted as a Caribbean destination, and one which is closer to many United States cities than more traditional resorts in Jamaica or Puerto Rico (Collins, 1979).

Existing facilities

In addition to any attractions and transport facilities, account must also be taken of the presence or absence of other plant, services and infrastructure as discussed in Ch. 2, notably: accommodation, shops, banks, health and security services, sewerage, power and water supplies. In some instances their absence will be a marked constraint on development; in others, their presence will represent an often under-utilized resource. A survey of tourism resources in Belize, for example, identified inadequate water supplies and effluent disposal systems as major limiting factors to tourist development while low occupancy rates indicated that the number of visitors to the main resort of San Pedro in 1982 could be doubled and still accommodated in existing hotels (Pearce, 1984).

Examination of existing facilities has primarily focused on the accommodation sector, with attention being paid to the distribution capacity and occupancy rates of particular forms, especially hotels.

Such an exercise may be an integral part of the planning process, where identification of pressures for expansion of accommodation may necessitate some policy response (Surrey County Council, 1984). In other instances, locational analysis of existing accommodation may indicate potential areas for development (S. Smith, 1977; Fesenmaier and Roehl, 1986). This procedure involves four basis steps:

1. Identification of a measure of accommodation development (the dependent variable) and a set of independent variables representing the tourism resource base (Smith) or describing the regional characteristics influencing accommodation development or success (Fesenmaier and Roehl).
2. Collection of these data in appropriate spatial units (e.g. counties).
3. Prediction of levels of accommodation development in each of these units through the use of such techniques as multiple regression analysis or automatic interaction detection.
4. Comparison of predicted and observed values to identify those which are currently saturated and offer little opportunity for further development and those which offer potential for expansion (observed values are significantly less than predicted ones).

Fesenmaier and Roehl found the existing distribution of campgrounds in Texas was strongly influenced by physical factors, particularly water in the form of lakes and reservoirs, proximity to population centres and access via major thoroughfares. An index of campground expansion potential was then derived for each county in Texas by comparing the predicted number of campgrounds for each (obtained by the multiple regression equation) to the actual number that existed. The results of this step indicated that a number of counties east and north of Houston had the highest potential for further development. Although concluding that this technique is flexible and cost-effective in providing useful information in the planning process, the authors were cautious in the interpretation of their results given (p. 22) that the analyses 'also indicated that many areas of the state exhibited unexpectedly high levels of development given their environmental setting, suggesting inadequacies in the model as well as identifying areas that are of unique character'. Smith (1977) also saw this technique as providing a valuable supplementary approach but concluded (p. 29): 'The prudent planner and entrepreneur will, of course, supplement this macro-level study with more detailed studies of site, economic, financial, real estate, market, transportation, tax, and other factors in areas that are of particular interest to him.'

Land tenure and use

The purchase of land or the acquisition of rights to occupy a site is a necessary prerequisite for any tourist development project.

The spontaneous development of second homes usually involves a series of individual initiatives on the part of both the buyers and sellers. Consequently, small plots of land are common in this type of development. In the case of large, planned projects and the creation *ex nihilo* of resorts, speedy acquisition of extensive tracts of land at the outset of the operation is important. Access to the required land gives the freedom to develop the resort as a whole, to construct it as an integral and functional unit according to specific architectural or town-planning principles. Rapid acquisition of the land is likely to minimize the effects of speculation, reduce legal costs and permit a more rapid return on any investment. Localities which offer large blocks of land in one or a few titles will therefore be favoured over those where holdings are small and fragmented or where titles are shared by many individuals. Variations in land ownership and tenure may not only influence the location of tourist resorts but also their form and, in certain cases, the developer himself (Pearce, 1979b). These may be especially important factors for tourism on small islands characterized by the limited availability of land and pressure on resources (Pearce, 1987a, pp. 160–2). The significance of the land factor in tourist development is emphasized by Davis and Simmons (1982, p. 216) in a review of World Bank supported tourism projects:

> Most frequently land acquisition and the problems associated with land acquisition were the principal factors causing delay. These delays had severe impacts on the costs of the projects during the years of great economic instability and high inflation in many countries, both in the developed and developing world. These factors were among the most important in reducing the economic returns on a number of projects below the levels anticipated at appraisal.

State land

New World mountain lands, in contrast to the long-settled alpine areas of Europe, are largely State controlled. Virtually all of the high country in New Zealand is Crown Land which is administered by different government departments. Although proceeding more cautiously at present, these departments have shown themselves to be reasonably receptive to ski-field development, with a variety of leases being given to both clubs and commercial developers (Pearce, 1978b), although in some cases concessions have not been granted. However, the conditions imposed by the lessors have influenced the form development has taken. Some park boards have resisted the opening of access roads. More importantly, no on-field commercial accommodation has been permitted, giving rise to a net separation between the ski-fields with uphill facilities and day shelters and nearby settlements which provide overnight accommodation. Coronet Peak and Queenstown are 19 km apart. A similar situation exists in Colorado (Thompson, 1971) where public land generally starts at the treeline and extends to the peaks. Uphill facilities have been permitted to be located there but the policies of the federal agencies have confined hotels, motels and cabins to the settlements in the valley bottoms and to other small enclaves of private land. Aspen and Vail, where this separation does not occur, are particularly successful resorts. However, as Simeral (1966) points out, leasing public land for the development of uphill facilities is generally less expensive than acquiring private land and may be particularly attractive where private tracts exist nearby on which to realize real estate operations.

The coast is another key tourism location where ownership is frequently vested in the State. In many countries the immediate foreshore is in the public domain though private ownership of adjacent land may effectively limit access to the beach. Many outstanding natural areas, particularly the more fragile ones, will also be in public ownership and protected through their status as natural or scenic reserves, and national and state parks. Depending on the status of such areas and public support for their designation, such lands may well be outside the developer's reach. In the mid-1980s, for example, the Club Méditerranée was unable to secure rights to develop a complex on a public reserve in Queenstown, New Zealand. Even the government's own Tourist Hotel Corporation was prevented from encroaching on reserved land there and as a consequence had to scale down their projected hotel. More generally, most national parks preclude private tourist dwellings and strictly control the amount and type of commercial accommodation and other visitor facilities. Elsewhere, development may be directed by certain standards and regulations as, for example, those regarding the development of Crown land for cottaging in Ontario (Priddle and Kreutzwiser, 1977, p. 171):

> The policy specifies that a minimum of 25 per cent of the shoreline must be retained by the province for public access and use, and that where there is currently less than 25 per cent public ownership, no further cottage development on Crown land will take place. No cottage development will occur on islands of less than 3 acres (1.2 ha), on lakes of less than 100 acres (40 ha). . . . Design standards are also provided which require buildings to be set back 66 ft (20 m) from the high water line, a minimum lot width of 150 ft (46 m) and a minimum depth of 400 ft (122 m). Furthermore, all lots must have a view of the water.

Communal land

In the French Alps, the high mountain lands are traditionally communal pastures whereas the better land around the villages tends to consist of small, fragmented freehold properties. Administration of the communal pastures is usually vested in the local municipality and a promoter need deal only with a single body, often only the mayor himself, to obtain rights to these. Usually a contract is signed between the two, making available a small amount of land to be sold freehold for the construction of the resort itself, with a 30-year concession of the ski-field (Perrin, 1971). This facility in obtaining rights to the communal pastures has been a significant factor contributing to the location of many of the new integrated resorts high above the level of traditional settlement (Reffay, 1974) and to their coherent, functional form.

Communalism, however, does not necessarily mean agreement within the unit as Nayacakalou (1972) points out in a different cultural context, that of Fiji. The different rights to the land are unevenly distributed amongst the owning groups, often giving rise to disputes over how a piece of land should be dealt with. Despite these complications, the Native Land Trust Board has developed a policy for the allocation of Fijian land for tourist development whereby the land owners are given the right to acquire a minimum of 10 per cent of the shares in the company which proposes to develop the site. Most hotels in Fiji to date, however, appear to have been built on alienated land belonging to expatriates. In some cases a freehold title encouraged an owner to move into the tourist industry, in others, foreign investors sought scarce freehold land to enhance the security of their investment. Ownership or favourable purchase of freehold land emerged as the top-ranked factor of Britton's (1980b) survey of hotel location preferences in Fiji (Table 5.2).

Table 5.2 Ranking of hotel location preferences in Fiji (1977) (first three priorities of each hotel)

Location factor	Rank	Responses
Ownership or favourable purchase arrangements of freehold land	1	16
Proximity to natural attractions	2	13
Favourable land leasing arrangements	3	9
Proximity to other tourist attractions	3	9
Proximity to international airports	3	9
Access to major road network	6	6
Influence of government policy	7	4
Availability of supporting services	8	3
Hotels available for purchase/leasing	8	3
Proximity to major urban area	10	2
Proximity to labour supply	11	1

(*Source*: Britton, 1980b).

Private land

Renard (1972) has demonstrated that in parts of France the system of tenure and the size of holdings effectively explain the distribution of second homes. Two basic tenurial systems exist along the Talmondais littoral (Vendee). Scattered amongst the large holdings of from 30 to 60 ha belonging mainly to absentee landlords and farmed by tenants, or *metayers*, are smaller owner-occupied holdings, often of less than 10 ha and fragmented in many tiny plots clustered around the hamlets. It is these small plots which have been colonized by second-home owners, whereas almost without exception no tourist development has occurred on the large properties whose owners have preferred to retain their investment in the land. However, in north-western Sicily, where the Mafia are still active, Ciaccio (1975) notes that absentee landlords have been quick to subdivide their large estates for second homes in order to reinvest their immediate gains in the city. Other examples of the role of tenurial systems on private land have been cited in earlier chapters, particularly with regard to the Balearic Islands (Bisson, 1986) and the Mediterranean coast of Spain (Dumas, 1976).

The price of land will of course be another major consideration. Where tourism is competing with other land uses, such as agriculture, industry or urbanization, prices will be much higher than where the land in question is not currently exploited, as may be the case with certain rural, upland or coastal areas. In urban areas in particular, different types of development will be able to support different land rents. More intensive forms such as high-rise hotels will be able to afford inner-city sites, whereas camping grounds will be forced to locate on the outskirts. At Waikiki in Honolulu, changing land values have fuelled the rapid growth of tourism and exacerbated the concentration of hotels and condominiums (Farrell, 1982).

Although the decision to sell or retain one's property is generally a private matter, public intervention may force the decision in favour of tourist development. In France, if a promoter can show his project to be in the public interest, the authorities may expropriate private land on his behalf (Baretje and Defert, 1972). Such powers do not exist in Switzerland where private property remains sacred. As a result, individual chalets may be found on the ski-slopes themselves, for example, at Verbier. In the case of the Languedoc–Roussillon littoral, the French government effectively implemented various land laws and regulations (ZAD (*zone d'aménagement différé*) and DUP (*déclaration d'utilité publique*)) to freeze the price of private land and give the State priority in land purchases. These measures enabled the State to limit speculators and acquire reasonably large and coherent blocks of land relatively quickly and cheaply ensuring the viability of the whole operation. In the Bahamas, the mechanism for the rapid growth of Freeport was the 1955 Hawkskill Creek Act which enabled an American to develop a huge block of land, attracting foreign investment by generous tax freedoms (Bounds, 1978).

Other considerations

Government policies other than those relating to land use may also be important factors to take into account. To encourage regional development, governments may make available various development incentives (low interest loans, tax exemptions, employment subsidies, developmental grants) in some regions and not others, or selectively facilitate tourist development through provision of infrastructure or promotional assistance. For instance, only those regions and areas within the European Community eligible for regional aid from national sources can be assisted by the European Regional Development Fund (Pearce, 1988a). Internationally, variations in public support from one country to another may be a decisive locational influence (Mings, 1978b). Spatial variations in the availability and cost of labour, the attitudes of residents to tourism and political stability will be other factors to bear in mind.

Carrying capacity

As Piperoglou (1967) has noted, before suitable regions or localities are developed they must be checked for their capacity to absorb tourists and new facilities and activities. Carrying capacity is commonly considered as the threshold of tourist activity beyond which facilities are saturated (physical capacity), the environment is degraded (environmental capacity), or visitor enjoyment is diminished (perceptual or psychological capacity). Lindsay (1986), with respect to national parks in the United States, suggests: 'Carrying capacity can be further defined by the formula:

$$CC = f(Q, T, N, Ut, DM, AB)$$

which states that carrying capacity is a function of the quantity of the park's resources, the tolerance of its resources to use, the number of visitors, the type of use, the design and management of the visitor facilities and the attitude and behaviour of its visitors and managers.' While generally seen as a useful planning and management concept, the establishment of levels of carrying capacity is no easy task (Barkham, 1973; Yapp and Barrow, 1979; WTO, 1984).

Many carrying capacity issues, particularly the complexity of the problem, can be seen with regard to coastal tourism (Pearce and Kirk, 1986). Much of the difficulty in assessing carrying capacities for coastal tourism stems from the composite character of tourism, the complex and dynamic nature of coastal systems and the different types of capacity which might be defined there. The major relationships for a typical coastal resort are depicted schematically in Fig. 5.3 which shows that different types of carrying capacity relate to and have been applied to different and often quite specific aspects of tourism and the coastal environment.

Dune systems are one of the more vulnerable, yet often overlooked parts of the coast. As Usher, Pitt and de Boer (1974) noted in their study of a Yorkshire coastal reserve, although the main attraction, the beaches, were themselves resistant to pressure, this was not the case with the dunes which first had to be crossed. Foredunes in particular are frequently subjected to great stress from intense use, tracking, vegetation depletion by pedestrian traffic and off-road vehicles, and 'landscaping' for building sites and car parks (Plate 10). Coastal dunes are delicate and their stability is easily disrupted; yet they are resilient, readily managed systems if sufficient attention is paid to their role in the coastal ecosystem (Pearce and Kirk, 1986).

Other environmental carrying capacity studies have dealt with offshore features and coastal waters. Salm (1986), for example, identifies the determinants of coral reef carrying capacity for underwater tourism as: the size and shape of the reef, the composition of the coral communities, the type of activities undertaken and the level of experience of the snorkellers and divers. Of more widespread concern

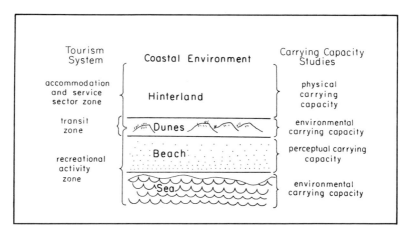

Fig. 5.3 Schematic representation of the spatial relationships between elements of the tourism system, the coastal environment and carrying capacity studies.
(*Source:* Pearce and Kirk, 1986).

Plate 10 Rather severe measures have been taken at Westhoek, Belgium, to discourage trampling of the dunes as holidaymakers move back and forth between the apartments and the beach. The question might be asked, however, whether such intensive development was appropriate in the first place

is the capacity of coastal waters to absorb discharged effluent. The Mediterranean, a major tourist destination, is particularly susceptible to degradation on account of its small tidal range and long water renewal period. The environmental carrying capacity of parts of the Mediterranean and other seas might be considered to have been surpassed when certain thresholds have been breached. As a WTO–UNEP (1983) report notes:

> From the human health point of view, tourism planners can refer to recommended bacteriological standards to ensure adequate protection for the water. For example, the Common Market Bathing Water Directive of 1976 set bacteriological limits for traditional bathing areas: 100 ml of sea water should not contain more than 10,000 coliform bacteria or 2,000 fecal coliforms.

A 1978 survey of 1200 beaches in France showed only 70 per cent had an acceptable water quality (Michaud, 1983).

In these cases, the breaching of the environmental carrying capacity is in large part a direct consequence of aspects of the resorts' physical capacity being surpassed, notably the capacity of their sewerage systems. Difficulties arise here not so much in assessing the appropriate capacity for local standards and design requirements are generally available (Salvato, 1976), but in delivering the necessary plant. The pronounced peaking of tourist demand along the coast exacerbates effluent discharge problems with many traditional communities experiencing tourist development being reluctant or unable to cope with pressures felt during only a 2 or 3 month period. The capacity of public sewerage treatment systems in France's 930 coastal communes was reported in 1979 to be able to cope with the needs of about 7 million people. This was sufficient for the winter population of about 6 million but clearly inadequate for the summer peak which sees this population double. Nevertheless some progress had been made, with the number of coastal sewage treatment plants increasing from 170 to 450 in the 10 years from 1970 (Michaud, 1983).

The increase in the number of sewerage systems is one example of how the physical capacity of infrastructure or plant may change. Another is the introduction of floating pontoons which has greatly increased the number of pleasure craft which can be moored in a given port or marina compared to traditional jetties. The continuing intensification of construction in resorts such as Torremolinos on Spain's Costa del Sol shows it is difficult to fix specific limits on the physical capacity of the accommodation zone. The amount of building undertaken on any site may be subject to some physical constraints but will also depend on a range of factors such as the technology available, the economics of the operation and aesthetic considerations (or the lack of them).

Developments at Torremolinos and in other resorts with very

high-rise sea fronts suggest that certain groups of tourists are prepared to tolerate remarkably high densities to get a place in the sun. Little work has been done in respect to the accommodation zone with most perceptual carrying capacities focusing on tourists' use of the beach and on determining an acceptable number of users for any given area. Recommended densities found in the 1960s literature range from 5 m² to 25 m² per user (Andric *et al.*, 1962; ACAU, 1967; Piperoglou, 1967; Service d'Etude d'Aménagement Touristique du Littoral, n.d.) though clearly far higher densities are tolerated, perhaps even sought, by the gregarious tourists who flock to the beaches of the Riviera or Costa del Sol. Observed densities on metropolitan beaches such as Ipanema in Rio de Janeiro are certainly often much higher although in many such instances little choice may be available to the local citizens who would get greater enjoyment from less crowded beaches if available.

Attempts at assessing the perceptual carrying capacity of beaches have involved relating the density of users to a measure of user satisfaction. One of the more useful methodological studies is that carried out at Brittas Bay in Ireland (An Foras Forbatha, 1973). Actual densities and distributions were obtained from aerial photographs taken on a Sunday in midsummer. At the same time, the views of the beach users were sought in a questionnaire survey. Comparison of the two surveys suggested that many users would accept a density of 1000 persons per hectare or 10 m² per person without considering a beach overcrowded.

Estimates of the perceptual carrying capacity of the beach can provide a useful input into the resort planning process, with the amount of accommodation (part of the physical capacity) being designed to limit crowding on the beach. Aerial photographs of developed beaches, for example, were used in planning the scale of the new coastal resorts in Languedoc–Roussillon but this approach has yet to be adopted more generally. In other instances, such as at Cancun and La Manga del Mar Menor, a conscious attempt to increase the amount of beach available appears to have been made by selecting coastal bar sites which have two waterfronts, one along the sea, the other facing inland to a large lagoon.

This review of carrying capacity for coastal tourism does not suggest the assessment of carrying capacity is a simple matter, with quantifiable thresholds which can be readily identified and incorporated into local plans or planning regulations. In some instances, as with the European Community's water quality standards, ready measures of environmental degradation do exist and the solution is apparent (more and better effluent treatment) even if the means of implementing this (more money) are not always immediately available. In other cases – for example damage to dune systems – it may, however, be possible to adopt management strategies to alleviate particular problems even if precise capacities have not been quantified (Pearce and Kirk, 1986). The notion that there may be limits to development beyond which the visitors will become dissatisfied, saturation may occur or the environment become degraded, may in itself lead to a more careful evaluation and exploitation of coastal and other resources.

One of the biggest challenges to those planning and managing the coast for tourism in the future will be drawing together the different types of carrying capacity depicted in Fig. 5.3. Establishing a level of recreational use purely in terms of tourist satisfaction will be inadequate if, as is often the case, this use is still excessive in terms of the dunes' resilience to erosion and of the sewerage system's capacity to treat effluent adequately. A more interdisciplinary approach to planning is needed to achieve this comprehensiveness.

More attention, in coastal and other areas, also needs to be directed at the social environment, that is at the capacity of the host communities to absorb tourists. In other words, attention needs to be paid not only to the social interaction between tourists but the relationship between them and the local residents. Cooke (1982, p. 23) and d'Amore (1983, p. 144) define community or social carrying capacity as 'that point in the growth of tourism where local residents perceive, on balance, an unacceptable level of social disbenefits from tourist development'. Both authors report on case studies in British Columbia which were undertaken to help form tourism policy. D'Amore, in particular, notes the complexities of tourist/resident interactions and the difficulty of quantifying such relationships beyond basic factors such as the ratio of tourists to residents. From a more subjective approach involving small group interviews in selected communities it was, however, possible to establish aspects of social sensitivity to tourism and detect variations from community to community in acceptable levels of development. Variations were due to differences in the types of community and levels and types of tourism involved.

Cooke and d'Amore subsequently propose a set of broadly similar 'guidelines for socially sensitive development'. Those put forward by Cooke (pp. 26–7) are:

1. At the local level, tourist planning should be based on overall development goals and priorities identified by residents.
2. The promotion of local attractions should be subject to resident endorsement.
3. The involvement of native people in the tourism industry in British Columbia should proceed only where the band considers that the integrity of their traditions and lifestyles will be respected.
4. Opportunities should be provided to obtain broad-based community participation in tourist events and activities.
5. Attempts to mitigate general growth problems identified in a given community should precede the introduction of tourism or any increase in existing levels of tourist activity.

Many of the processes and examples discussed in Ch. 3 would suggest that such considerations are frequently neglected, with external developers and their clients (foreign or extra-regional tourists) dominating local residents and their needs. Other aspects of social impact and associated planning issues are discussed in subsequent chapters.

SITE SELECTION AND REGIONAL RESOURCE EVALUATION TECHNIQUES

Two similar but distinct areas of planning and decision-making will involve a spatial differentiation of the factors discussed here. Firstly, it may be necessary to determine priorities for development amongst a given range of sites, that is, it is a question of site selection (Gearing and Var, 1977). Secondly, general evaluations for regional or national planning will require the identification and delimitation of those areas most suitable for one or more forms of tourist development (Piperoglou, 1967; Georgulas, 1970; Vedenin and Miroschnichenko, 1970; Lawson and Baud-Bovy, 1977; Var, Beck and Loftus, 1977; Carvajal and Patri, 1979; Gunn, 1979).

Similar methodologies have been used in each case. These have usually involved the following basic steps:

1. Selection and possible weighting of criteria.
2. Evaluation of sites or areas in terms of these.
3. Derivation of a relative measure of overall potential or attractiveness.

Few, if any, writers appear to have employed the full range of criteria discussed in the first section. Access and availability of land are the most frequently neglected. The land factor is of course very specific and may enter the decision-making process only after suitable broad areas have first been determined. Access, however, is far more critical. One of the few cases where access was emphasized was that of the Chilean Antarctic region (Carvajal and Patri, 1979). The criteria selected also depend on the specific objectives of the exercise. In their analysis of coastal locations in Turkey, Gearing and Var (1977) used six general factors – natural beauty, pollution level, historical value, distinctive local attractions, attitudes towards tourists, other recreative possibilities – and seven more specific ones affecting bathing – user density, water temperature, clarity and turbulence, wind factors, sand quality and beach descent. Different natural factors were used by Vedenin and Miroschnichenko (1970) in their recreational evaluation of Russia according to the season. For practical purposes, most of the writers have limited themselves to from ten to twenty factors. However, it may be possible to incorporate a larger number of factors in a general evaluation by first reducing some to a single common index, as with climate or attractions. Care must be taken here for, as Gearing and Var stress, the factors should be independent and the attractivity indices of Piperoglou (1967) and Ferrario (1979) already include access and infrastructural components.

Having selected appropriate criteria these must then be weighted, for it is unlikely that each will be of equal value. Surveys of tourist preferences aid both in the selection of criteria and their weighting (Piperoglou, 1967; Ferrario, 1979). Such surveys are generally costly and may be inappropriate for assessing all factors; for example, access, infrastructure and land availability. Recourse to 'expert evaluation' is the most common solution (Gearing and Var, 1977; Var *et al.*, 1977; Ritchie and Zins, 1978). Gunn (1979) simply assigns weights through arbitrary subjective selection. This is also probably the case for those others whose weighting procedures are not particularly clear. As well as this critical weighting of factors relative to each other, more attention could be given to a subsequent weighting based on the spatial coincidence of the various factors (Piperoglou, 1967; Ferrario, 1979).

A further question concerns the unit area of measurement for these criteria. Vedenin and Miroschnichenko (1970), for instance, employ the natural provinces of the USSR and Var *et al.* (1977) planning districts in British Columbia. The use of planning districts may be relevant where tourism is one of a number of development options but there is no inherent reason why tourist regions should correspond to any existing administrative units. The use of point data, where these are available and applicable, will therefore usually be the most appropriate, enabling tourist regions to be subsequently defined in terms of relevant parameters (Piperoglou, 1967; Ferrario, 1979).

Having determined the criteria which will discriminate between areas or sites with regard to their tourist potential, the regions or alternative sites in question must be evaluated in these terms. A basic approach is to rank them on some ordinal scale, not necessarily numerical; for instance, from most favourable to unfavourable (Vedenin and Miroschnichenko 1970; Gearing and Var, 1977). At this stage, certain regions or alternatives may be eliminated from further analysis through not attaining a designated threshold on key criteria; for example, climate or access. Gearing and Var also eliminate those sites which are 'dominated' by others, that is, those which consistently score lower on all criteria. To identify the most favourable regions, Vedenin and Miroschnichenko simply add the total number of points scored by each province on a five-point scale. Gearing and Var, however, employ a slightly more complicated 'quasi-lexicographic' procedure whereby alternatives scoring a maximum value on each criterion are given a score equal to that criterion's weight, the sum of these scores giving a composite score for each alternative. Other writers, however, score each alternative on every criterion used, not solely those attaining maximum values. Var *et al.* (1977) sought the co-operation of experts who were asked to evaluate districts with which they are familiar on a 0–100 scoring system. Average scores were multiplied by the criteria weights to give weighted total scores for each district. A similar approach is used by Carvajal and Patri (1979), in their study of the Chilean Antarctic region, only they appear to do their own evaluation using a much reduced scale (0–4), based mainly on the presence or absence of various conditions. As was noted earlier, Ferrario (1979) takes the number of mentions in guide books as a surrogate for expert evaluation of particular attractions.

Cartography is employed by other writers as an alternative to these essentially quantitative approaches (Lawson and Baud-Bovy, 1977; Gunn, 1979). This generally involves a thorough inventory of the region in question in terms of the criteria selected and the production of a corresponding series of maps. Synthesis may be achieved by the use of overlays to identify major tourist regions. Gunn, and Molnar and Tozsa (1984), however, suggest the use of computer mapping incorporating the weighting of different factors and their aggregation into composite scores (Fig. 5.4). But as Piperoglou (1967) has shown, this may also be achieved through use of more conventional techniques.

A further approach is suggested by Georgulas (1970). He relies on the establishment of a classification system of different destination points. A number of functional elements are first identified. These are ranked according to a series of specific criteria (e.g. for first-class beaches, see above). A detailed notation system is then used to present information about each feature in tabulated form and cartographically. The clustering of features on the maps enables the identification of tourist regions.

Clearly the evaluation of the tourist potential of an area is a complex task and one where methodologies could be further developed and refined. A key problem is the comparison and weighting of a wide range of parameters, for as Nefedova Smirnova and Schvidchenko (1974, p. 507) point out: 'Any attempt to apply mathematical techniques to these multifactor evaluations without a sound basis for factor weighting is bound to fail in principle. But when factors can be logically weighted, quantitative techniques do assume significance'.

However, before techniques get too sophisticated or complicated it should be kept in mind that for the purposes of planning and most decision-making it is the relative importance of one location to another rather than absolute values which are initially important. At a subsequent stage, once initial choices have been made, the costing and economic analysis of particular sites, projects or regions will re-enter the process (Gearing and Var, 1977). Given that few private developers and most

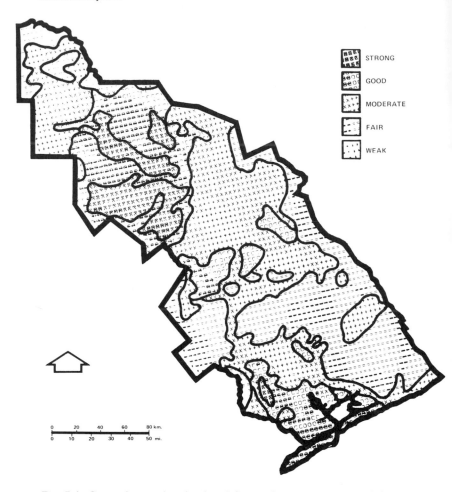

STRONG

GOOD

MODERATE

FAIR

WEAK

Fig. 5.4 Sum of natural and cultural factors for tourism potential (twenty counties in Texas).
(*Source:* Gunn, 1979).

public authorities will not commit large sums for such exercises at the outset of the development process, techniques which offer a ready yet accurate assessment and which are not over-demanding in terms of data, time or money will prove the most practical.

Lack of finance or willingness to commit funds may be only one reason why full-scale evaluations are not carried out. Where the initiative is a local or even a regional one, the choice is likely to be restricted to nearby sites and attractions and any research will be to determine the feasibility of developing these. In such cases, the resultant pattern of development may be more a reflection of variations in enterprise than of the other factors outlined.

The basis of locational decision-making may also change through time. As Robertson (1977, p. 122) points out in the case of second homes: 'The last buyers in a particular second-home sub-division face a set of site characteristics that is quite different from those faced by the first.' Similarly, someone wanting to open an *après-ski* facility will be interested not so much in the physical characteristics of the ski-field but in the size and nature of the market which the resort can provide.

Ski-field location

Ski-field location, perhaps more than any other type of tourism, is heavily dependent on appropriate physical conditions, namely a good snowcover and suitable slopes.

Two main factors are important with regard to snowcover – the length of the season and the reliability of the snowfall. Longer seasons (of from 4 to 6 months) and more reliable snowfalls will yield a more satisfactory and reliable return on investment in plant and infrastructure. Reliability will also build confidence amongst skiers. The duration of the snowcover is closely tied to altitude. At higher altitudes, say 2000 m, the snowfall is generally not only greater, but the snow remains longer as temperatures are lower. Aspect and exposure to winds will also determine the amount of snow retention and the need for snow-packing equipment. Duration of the snowcover is also a function of latitude; similar altitudes in the Pyrenees and the French Alps will experience different conditions. Martinelli (1976) stresses the importance of the early season snowcover as this is when demand is often the greatest and when the financial success or failure of many areas is decided. The quality of the snowcover is also important, with skiing performance being related to the following snow parameters: density, temperature, liquid water, hardness and texture (Perla and Glenne, 1981). Account must be taken too of other climatic factors. Wind conditions may not only influence retention of the snow but affect the functioning of lifts and the comfort of skiers. Likewise, temperature variations will affect the quality of the snow and the comfort of skiers as well as the design of buildings. Frequent fogs may lead to lengthy closures of the field. Sunshine is particularly important for the siting of resort accommodation but also a significant factor in skier enjoyment.

Resort ski-fields must offer a range of slopes to attract a variety of skiers. Sibley (1982) reports that novice slopes should be less than 20°, intermediate slopes, 20–45° and advanced slopes, 45–80°. These values are somewhat higher than those suggested earlier by Rey (1968), perhaps an indication of some evolution in general skiing ability. The number of trails available and the total vertical drop are also important

– experienced skiers will require a variety of slopes to ski during the day. Farwell (1970) proposes the total vertical drop be in the order of 1500 vertical feet in the North-eastern United States and from 3000 to 4000 vertical feet in the West. In France, the physical capacity of ski-fields is reckoned to be in the order of one skier per metre of vertical drop (Cumin, 1966). The capacity of a given field will depend in part on the range of skiers it attracts with the capacity of any lift being determined by 'multiplying its capacity per hour by its vertical rise and then dividing by the number of vertical feet per hour, a skier of the run's ability, served by the lift, can ski' (Sibley, 1982, p. 81). Procedures for calculating lift capacity in this way are outlined by Sibley. Sibley also notes a Canadian rule-of-thumb by which a ski area is considered to be at capacity when it reaches 40–45 per cent of total possible skier–days for the whole season. It is interesting to note that in her extensive study of three Aspen ski-fields, London (1975) found skiers' evaluations differed considerably from actual use statistics, suggesting marked differences between actual and perceived capacities. At the same time, a certain amount of flat stable land must be available close by to provide adequate building sites. Simeral (1966) suggests, as a general rule, 2 ha of base land for every 400 ha of hill operations, though this will depend on the form the resort takes (cf. La Plagne, Fig. 3.3). Considerable care must be taken in locating both the base and uphill facilities to avoid avalanche paths (Seligman, 1980). Avalanche conditions may severely limit the capacity of certain fields.

France's Service d'Étude de l'Aménagement de la Montagne suggests an 'ideal' ski area would be a cirque, offering a variety of *pistes*, in terms of aspect and difficulty. These *pistes* would converge on a single reception area where the accommodation and other facilities would be built. Such a form gives skiers immediate access to all parts of the ski-field. These conditions are approached in several of the new resorts such as La Plagne, Pra Loup and Tignes. Ski-fields set in valleys are not generally accessible from a single centre, for example, Vars. On the other hand, access to them is generally easier than to the cirques which may require costly access roads.

The suitability of the physical conditions found above 2000 m in the French Alps frequently coincides with the existence of communal pastures where rights to occupy the land can usually be obtained without too much difficulty. Elsewhere such sites may be on Crown or federal land where policies, while allowing development of lifts, may not permit the construction of accommodation and associated facilities.

The value of any one site will increase where the possibility of linkages with other ski-fields exist, for skiers, more than their coastal counterparts, seek variety on their holidays. A location within or near to an already well established and known ski area will be favoured over a more isolated one. Thus, Aime 2000 forms part of the La Grande Plagne complex which in turn is favourably situated in the Tarentaise. Proximity to summer attractions or resorts is another advantage. Real estate sales may benefit from this double season or the ski-field may profit from existing infrastructure and accommodation. Chamonix, for example, developed first as a summer resort and the summer traffic still exceeds that of the winter (Veyret-Verner, 1972).

Proximity to the market is another key factor. As Stansfield (1973, p. 6) points out in the case of New Jersey's ski industry; 'It is apparent that otherwise mediocre slopes in combination with hazardously unpredictable local climate conditions can, if accessible to population centres, easily be transformed into a successful ski recreation area.' Access and proximity to urban areas is especially important for those centres catering for day visitors and weekend skiers, as is the case with most Canterbury (New Zealand) ski-fields (Pearce, 1978b). Vacation resorts, on the other hand, may be more distant but will need to offer a wider range of facilities and generally better skiing conditions.

Several recent studies which have analysed in more detail aspects of skier demand using techniques discussed in Ch. 4 provide a useful complement to those which have focused directly on the physical characteristics of ski areas. In a detailed study of Grenoble skiers, Keogh (1980) shows 'sporting' skiers give more weight to the length of ski-runs in selecting their ski-field whereas 'social' skiers are more concerned by the cost of lift tickets. Ewing and Kulka (1979) found the attractiveness of Vermont ski resorts was largely perceived as a function of two sets of interrelated factors, perceived length and variety of slopes on the one hand and perceived crowding and price on the other. Boon (1984) analysed the benefits sought by skiers in North Carolina and discovered 'being with friends' was the most important outcome sought by the skiing experience. This finding contrasted with the results of a similar study of Lake Tahoe area downhill skiers where exercise and physical fitness were seen as the most important, leading Boon (p. 399) to conclude that 'southern skiers differ from their western and midwestern counterparts in terms of overall benefits sought'. Boon also reports that advertising specifically tailored to the market identified by the study led to increased skier visitation.

CONCLUSIONS

A wide variety of factors must be taken into account when evaluating the tourist potential of any area. The importance of each of these will vary from situation to situation depending on the type of tourism being developed, the context and stage of development and the perspective adopted – that of the developer, the tourist or the host community.

Even in the more specific case of ski-fields, different factors assume different emphasis according to the location and type of skier being catered for. However, the range of evaluation and selection techniques reviewed do provide a good basis for assessing these and ordering these factors although there is still scope for refining these techniques further. In particular, further consideration needs to be given to assessing the conditions conducive to meeting basic motivations of the need to escape and to rest and relax and to evaluating the less tangible cultural attributes attractive to wanderlust tourists. Destination features need to be seen more directly in terms of market characteristics and tourist motivations. At the same time, more attention should be paid to the capacity of destinations to absorb tourists in terms of the interrelated types of carrying capacity discussed earlier. Evaluation of tourist resources in this wider sense will normally require a broader interdisciplinary approach to the problem. The skills needed, for example, to evaluate the snow conditions of a particular basin, to identify the appropriate segments of the skier market and to assess the community carrying capacity of associated settlements are unlikely to be found in a single person or discipline.

6

Analysing the impact of tourist development

The impact of tourism was identified as the most common theme in the review of tourism research presented in Ch. 1, with varying emphasis being given to economic, social/cultural and environmental issues depending on the writers' background and discipline. It was also suggested later in Ch. 1 that many such impact studies failed to consider adequately the type of tourism concerned and that impacts were often divorced from the processes which had caused them. Moreover, studies of the impacts of tourism are not usually set in any broader context of development. Preceding chapters have also shown that different concepts of what constitutes development exist and that different processes of tourist development occur. In this chapter it is argued that the varied impacts of tourism reflect the context and processes of tourist development and that these condition tourism's contribution to the state of development in any area.

This chapter begins with a review of two general frameworks aimed at structuring the wide variety of impacts which tourism may generate. Discussion then focuses in turn on the three generally recognized subsets of impacts: economic, social/cultural and environmental. In each of these sections, the discussion is structured around a specific framework depicting the range of impacts which are to be found and the relationships between them. Consideration in these sections is also given to methodological questions, for the nature and extent of the impacts identified in the literature commonly depends heavily on the techniques employed to assess them, and to relating impacts to particular processes of tourist development.

The order the sub-sets of impacts are dealt with broadly follows the way they have emerged as issues in the impact literature and the direction the broader debate on development has taken. The focus in the 1960s was largely on economic matters, with social/cultural and environmental concerns being voiced increasingly throughout the 1970s, in part a consequence of the emergence of a more general awareness of these issues, particularly environmental ones (Krapf, 1961; Kassé, 1973; Bryden, 1973; V. Smith, 1977b; de Kadt, 1979; Mabogunje, 1980). By the early 1980s there was a general recognition in the impact literature that all three sub-sets of impacts should be taken into account (even if this was rarely done) and that both positive and negative impacts could be attributed to tourism (Mathieson and Wall, 1982; Murphy, 1985). Meanwhile, the development literature continued to evolve along the diverse paths outlined in Ch. 1, with, for example, the various adverse environmental and social impacts being recognized as externalities of development.

GENERAL FRAMEWORKS FOR IMPACT ASSESSMENT

While most studies on the impact of tourism have been concerned with development which has already occurred, impact assessment is now also included in some planning processes. With the emergence during the 1970s of environmental impact auditing procedures in the United States, New Zealand and elsewhere, a new kind of impact assessment emerged, that of appraising the likely effects of development before a project goes ahead (Morgan, 1983). As Phillips (1974, p. 63) noted: 'in theory at least, the burden of proof now falls on the person or group wishing to disturb the environment. It is necessary for them to show that the proposed action will not impair environmental quality or that the social benefits of the action will outweigh the social costs.' In New Zealand, for example, environmental impact reporting and auditing procedures played key roles in the proposal to develop the Rastus Burn Ski-field in the Remarkables near Queenstown, with stringent conditions subsequently being imposed on the construction of the access road (Sibley, 1982).

Potter (1978) presents a very useful methodology for impact assessment, based partly on the experience of analysing the impact of oil drilling platform construction in Scotland. Potter's methodology, when broadened and modified to consider not only environmental but also social and economic impacts, provides an extremely good framework for investigating the impact of tourist development. Moreover, it can be usefully applied to both future projects and those which have already

Table 6.1   A general framework for assessing the impact of tourism.

1.   Examine context – environment, society, economy.
2.   Forecast future if tourist development does not proceed/had not proceeded.
3.   Examine tourist development.
4.   Forecast future if development proceeds/examine what happened when development occurred.
5.   Identify in quantitative and qualitative terms differences between 2 and 4.
6.   Suggest amelioration measures to reduce adverse impacts.
7.   Analyse the impacts and compare alternatives.
8.   Present the results.
9.   Make a decision.

(*Source*: after Potter, 1978).

taken place (Table 6.1). The basic steps (1–9) described below are involved.

1. Examine context – environment, society, economy

The first step is to examine the context of development, either that which existed prior to the advent of tourism or the contemporary situation where the concern is with the study of potential impact (Ch. 2).

Physical environment

In order to gauge its resiliency and suitability for development, the environment must be examined in terms of its physical characteristics – soil, vegetation, relief, aspect, fauna, climate – and the dynamics of the relationships between these. Attention, for example, must be paid to slope or dune stability, run-off characteristics and the resiliency of biotic communities. As well as a general environmental assessment, micro-level studies must also be undertaken, for example, determining the precise location of avalanche corridors. In the case of a built environment, attention must be directed at the nature, scale, form and location of existing buildings and street patterns as well as existing land uses.

Society

Characteristics of the host society to be taken into account include population size; demographic composition and vitality; ethnic, social or religious structure. For example, is the population increasing or decreasing, is it composed of more than one ethnic group, what social customs are evident? As White (1974, pp. 35–6) notes, such social/demographic characteristics can have a significant influence on the degree of development and change: 'a strong area can sustain the capitalisation and provision of labour from within itself, while a weak area is immediately more susceptible to outside economic influences in the form of external investment and immigration, so that the local socio-cultural structure is quickly changed'.

Economy

The size, diversity and vitality of the economy at various scales – national, regional or local – are further factors which need to be considered. The economy in question may be well developed, developing, depressed or in decline. It may be broadly based or heavily reliant on a single sector, a strong economy or a dependent one. Tourism may already be a significant sector or totally unimportant. These factors will influence the extent of local participation and degree of external involvement as well as determine the costs and benefits to different sectors of society and the ultimate impact of tourism.

2. Forecast future if tourist development does not proceed/had not proceeded

Having examined the existing situation and having established contemporary trends, the next step is to predict what would happen/would have happened if tourism does not develop/had not developed. This may involve a projection of demographic and economic trends. What would have happened, for example, to the alpine communities discussed in Ch. 3 had tourism not been developed?

3. Examine tourist development

As has been stressed earlier, it is essential not to think of tourist development as merely the growth of some monolithic phenomenon known simply as 'tourism.' Each particular project or development must be examined in terms of the range of specific elements and the process characteristics outlined in Chs. 2 and 3. What precisely is being developed, by whom, for whom, how, when, where and why are the fundamental questions which must be asked here. The question of timing, for example, is critical. Where tourism develops gradually over a number of years there will be a longer period for social and environmental adjustment and local participation may be increased as needs in terms of labour, supplies and capital will be met more readily by local resources. Where rapid development occurs, there may be apparently greater immediate returns but it will be unlikely that the host community can respond to such demands and provide all that is needed so that, in the medium and long term, 'leakages' from the area may be large, effectively reducing benefits to the community.

4. Forecast future if development proceeds/examine what resulted when development occurred

What parts of the environment, segments of society and sectors of the economy will be affected/have been affected, by what (specifically), in what ways and to what extent are the principal questions to be asked next. The impacts of past developments will usually be assessed more readily than future impacts can be predicted, particularly if the conditions existing before the advent of tourism have been satisfactorily established. Comparison with past developments will also be useful in assessing future impacts, keeping in mind, of course, the effect of differing contexts and types of development.

5. Identify in quantitative and qualitative terms differences between steps 2 and 4

To be able to evaluate fully the impact of tourist development the difference between tourism developing and it not developing must be assessed, that is, the difference between the results of steps 2 and 4. This involves measurement of a wide range of parameters in quantitative terms and a more general qualitative assessment of others. What should be examined, and how, are questions which have been subject to a certain amount of debate and answered with varying degrees of success. Two associated problems arise – the techniques to be used and the availability of appropriate data and information. Data deficiencies are very common, particularly in studies of past developments, for as Figuerola (1976) points out, there is often a shortage of even the most basic statistics. In general there are few earlier reports to draw upon or with which to compare the present. A further difficulty arises here in comparing the different types of impact. Whereas economic costs and benefits may be presented in purely dollar terms, social and environmental issues cannot usually be expressed so conveniently. Problems may also arise in establishing the causal links between different facets of development and changes in the host society or environment. It might be shown rather readily that new employment opportunities have arisen out of the development of tourist facilities but accompanying changes in the economic aspirations and behaviour of these workers may be due as much to external influences (such as the mass media) as to direct experience of the tourist industry. Various techniques used for examining the different impacts of tourism are described more fully below.

6. Suggest amelioration measures to reduce adverse impacts

Having identified sources of adverse impact, various measures might be suggested to reduce these. Spatially, for example, two broad strategies might be adopted – concentration or dispersal of tourist activity (see Ch. 7). The seasonal concentration of tourism is a further problem and a staggering of holidays may alleviate social and environmental pressures and lead to a more balanced economy. More specific suggestions might be to relocate facilities, increase the capacity of sewage treatment plants or to limit the activities of visitors through zoning.

7. Analyse the impacts and compare alternatives (where available)

Where alternatives exist, the merits of each case should be assessed and their impacts compared. Basically, two sets of alternatives may be identified. Firstly, there may be alternative sites available for a specific project, for example, the development of a new ski-field. Is site A more environmentally stable than site B? Secondly, in a particular context, a range of development possibilities may exist. Overall, is project X more desirable than project Y? In addition, it may also be necessary to compare the impact of tourism with other development strategies if a range of choices exists. What will be/would have been the impact of developing forestry or expanding agriculture instead of tourism?

8. Present the results

The results of the various studies should then be presented as clearly, concisely and objectively as possible. In the interests of a broad audience and a more ready comprehension of the results, technical jargon, complex formulae and the like should be reduced to a minimum or confined to appendices.

9. Make a decision

In practice, many studies, particularly of past developments, have ignored several of these steps. Attention has commonly focused on step 4, what resulted when development occurred with only cursory discussion of the context and project and little or no consideration of what would have happened had there been no tourist development. Moreover, few impact studies are at all comprehensive, with most concentrating on a single domain be it economic, environmental or social/cultural. Factors accounting for the incompleteness of past studies include: data deficiencies, inadequate methodologies, lack of resources, a failure to appreciate the different processes of development, the belated recognition of issues other than economic ones, the scale of analysis and the lack of an interdisciplinary approach to the problem – most impact studies have been written by one or two researchers from the same discipline, though the assessment of future impacts is more often undertaken on a broader basis.

There has also been a tendency for research on the impact of tourist development to concentrate solely on the destination areas. This is in part an outcome of the lack of emphasis given to the processes of development and a failure to recognize the linkages between origins and destinations or demand and supply, both geographically and structurally (Thurot, 1980; Pearce, 1987a). Thurot adopts a more holistic approach in his review of carrying capacity and presents a model in which he distinguishes between internal (contextual) factors in the destination region associated with the host society and economy and the local environment, and external ones, whose origins lie elsewhere, either in the generating region or the transit region (Fig. 6.1). In particular, conditions in the generating region will not only affect total demand, but factors such as institutionalized holidaymaking and national habits and attitudes may give rise to a marked seasonality in demand (Fig. 4.2) and to distinct preferences for certain types of holidays and destinations (Ch. 4). These conditions can accentuate the impact felt in the destination region by concentrating demand spatially and at certain times of the year, week or day. How the tourist industry is organized may also reinforce or reduce these patterns, with much domestic tourism being unstructured and more dispersed and international tourism being more organized and directed. The nature, capacity and organization of the transport network in the transit region may accentuate further this spatial and temporal concentration. Conditions in the transit region, notably those which influence the cost and ease of travel, may also significantly affect the volume of traffic generated and thus the pressure which develops at the destination. This region will also be affected by demand emanating from the generating region. Severe bottlenecks may be experienced at the interfaces of the different regions; that is, the outskirts of cities

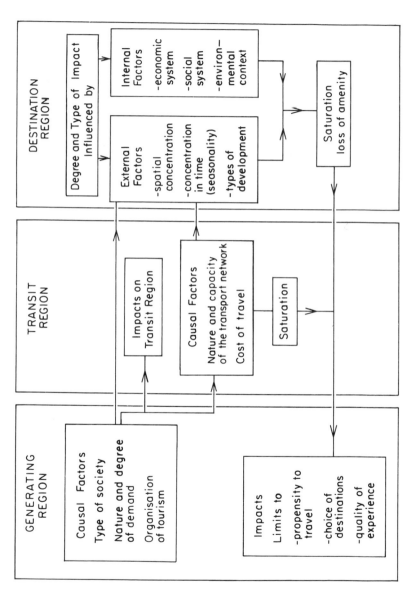

Fig. 6.1 Interrelationships between causal factors and impacts associated with generating, transit and destination regions.

and the approaches to resorts, at the beginning and end of the main holiday season and at the start and finish of the weekend. Moreover, feedback loops might be built into the system. If the destination region becomes saturated, this will limit the propensity to travel from the generating region, reduce the amount of choice available to potential holidaymakers or lower the quality of their experience such that the original need to escape may not be satisfied. However, the impacts felt in the generating region need not necessarily be all negative, particularly if economic considerations are taken into account. Significant income and employment may be generated there as a consequence of travel to other regions. This overlooked and underevaluated point is discussed further in the following section.

With its emphasis on the interrelationships between causal factors, and impacts associated with generating, transit and destination regions, Fig. 6.1 provides a valuable complement to Table 6.1 as a general framework for assessing the impact of tourist development. In addition to reinforcing the importance of contextual factors (step 1 in Potter's framework), the broader perspective depicted in Fig. 6.1 can shed further light on the type of tourist development under consideration (step 3), perhaps suggest a greater variety of amelioration measures (step 6) as well as indicate a fuller range of affected regions and types of impacts.

While some of these relationships have already been identified and explored to some extent elsewhere, such as the core–periphery linkages depicted in Fig. 3.7, the approach represented by Fig. 6.1 has not yet been generally adopted in the tourist impact literature. One of the few impact studies incorporating processes of spatial interaction is that by an interdisciplinary team concerned with assessing environmental stress arising out of the growth of ski tourism in Switzerland (Federal Department of Forestry, 1981). A central feature of their study was a model of ski tourism and the natural landscape comprised of three sub-systems: A – the outward and return trip, B – accommodation and stay in the resort, C – skiing and associated facilities. Each of these sub-systems was then examined in more detail in terms of the specific problems which arose and possible solutions which might be implemented. Problems in sub-system A, for instance, essentially result from the traffic generated by skiers and give rise to such impacts as an overloading of arterial routes, congestion in villages *en route* to the ski-fields, bottlenecks in passes, increased vehicle emissions and so forth. Solutions to these problems include extending existing roads or building new ones, creating a more efficient public transport system or developing alternative ski-fields and redirecting traffic to them. Each of these choices may ease the pressure in one area but generate new skiers elsewhere. Bypasses, for instance, may relieve congestion caused by the transit traffic in the villages but at the same time directly result in the loss of cultivated land and, by improving access, attract additional traffic which may cause greater emissions and increase demand on the other two sub-systems.

ESTIMATING THE ECONOMIC IMPACT OF TOURIST DEVELOPMENT

The objective and detailed evaluation of the economic impact of tourism can be a long and complicated task. Much of the complexity arises out of the disparate nature of the industry and the range of benefits and costs which can accrue to or are borne by various groups or individuals. Figure 6.2 attempts to provide a general framework by which to approach this problem by identifying systematically the different issues and relationships involved. For either the development or operational phase, the emphasis is on relating particular costs and benefits to specific aspects of development and to particular groups of people.

Development versus operation

It is first of all useful to distinguish between the developmental stage, that is when plant and facilities are being constructed, and the operational stage, when they are being used by the visitor. In the first phase, most of the capital will be that of the developers (or their financiers). In the second, it will largely be tourist expenditure which is circulating in the economy. The transition from one stage to the next may be accompanied by differing costs and benefits. Much of the local impact from second homes comes from the initial sale of land with little subsequent job creation or spending in the area by the second-home owners (Barbier, 1977). State expenditure on infrastructure may initially be very heavy but some expenses will be recuperated in the long term, for example, through increased tax revenues (Fernandez Fuster, 1974). Thus the stage at or time period during which a project is examined may influence significantly its economic evaluation.

The spectrum of development

Any comprehensive economic impact study must consider the whole spectrum of the tourist project or national and regional tourist industry. It is insufficient to concentrate on the accommodation sector alone, even if it is commonly the single largest element of investment or tourist expenditure. Account must also be taken of the other sectors shown in Fig. 6.2 and discussed in Ch. 2. Patterns of ownership, investment costs, income and employment generation will vary from sector to sector and each of these needs to be taken into consideration to get a full picture. Where the focus is on the impact in the destination, as it usually is, it may be legitimate to ignore the impact of market-oriented services in the origins, even if they can be conceptualized as being generated by the same acts of tourism. This may be particularly the case with

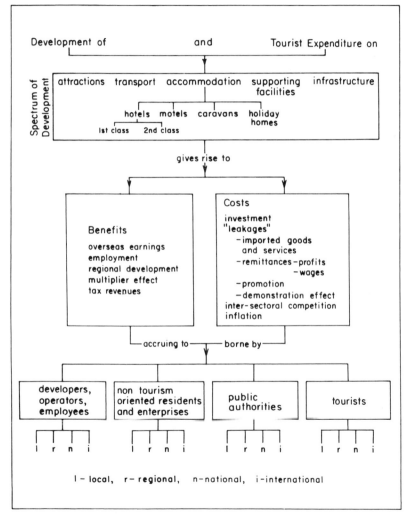

Fig. 6.2 A framework for assessing the economic impact of tourist development.

developing countries which generate little travel abroad. However, in developed countries with large volumes of outbound as well as inbound traffic, the effects of these market-oriented services may be significant as Table 2.2 indicated. France's 2300 travel agencies, for example, had 18,000 employees in 1983. This represented 3 per cent of all employment in a very broadly defined tourist industry (WTO, 1986). Within each sector it is also necessary to take account of the particular type of development, for instance to distinguish between the different types of accommodation and their associated impacts.

Major technical problems, however, arise in the actual delimitation of the spectrum of tourist development for purposes of economic analysis (WTO, 1980b, 1983b). In most cases, tourism is defined in terms of the composite demand for a range of goods and services sought by tourists. In this respect, tourism differs from other activities such as agriculture and industry which 'are determined in terms of the goods and services generated by units of production' (WTO, 1983b, p. 48). In practice, most national accounting systems do not have a specific tourism category so that 'rather than an economic sector, tourism activity has to be seen as a grouping of conventionally defined sectors' (WTO, 1980b, p. 6). However, even in apparently sharply defined sectors such as hotels, not all the income or employment can be attributed to tourist demand, for example that resulting from bar sales to local residents, a major component of hotel business in urban areas. While some researchers may deflate their figures accordingly, often such considerations do not appear to be taken into account. To avoid these problems tourism is often delimited by tourist expenditure on different goods and services, that is in terms of demand (Pearce, 1981). Different practices in delimiting tourism can clearly have significant effects on the figures produced.

Benefits and costs

The main economic benefits which tourist development may bring were outlined in Ch. 1 in terms of reasons for advocating tourism as a path for economic development and in Ch. 2 in terms of the motives the public and private sectors have for developing tourism. These benefits are reduced or modified by a series of associated costs, with the final economic impact of tourism resulting from the interplay and balance of both sets of factors.

Balance of payments

At a national level, the major aim of both developed and developing countries in promoting tourism is commonly to increase foreign exchange earnings, to improve or redress the balance of payments situation. This is also true of many East European countries, some of which even impose minimum currency exchange regulations on foreign visitors (Tassin, 1984). Development of international tourism offers three main advantages for increasing foreign exchange earnings:

1. It has been, and still is, a growth industry (Fig. 4.3).
2. The tourism market, unlike that for many manufactured or primary goods, is relatively little protected. Moreover it is a market which comes to the producer.
3. For many countries, tourism may represent a diversification of the economy and a means of lessening reliance on traditional exports.

Travel expenditure is classified in conventional balance of payments accounts as a separate item under 'invisibles', and is usually estimated from bank transfers, travel agency records or tourist expenditure surveys. However, as White and Walker (1982) point out, these are frequently 'very rough' estimates. Different practices and definitions have been adopted by different countries (e.g. transportation is included as a separate item or combined in varying degrees with travel or different definitions of tourists and residents are used). Comparative studies reveal often quite significant inconsistencies from country to country. Published foreign exchange figures relating to tourism must therefore be treated circumspectly.

Tourism's role in the balance of payments is commonly expressed in terms of revenue generated by international visitors and expenditure by nationals abroad, both in absolute and relative terms along the lines shown in Table 6.2. This table shows that Italy, Spain and France are the top three member states of the European Community in terms of absolute earnings from international tourism. In relative terms, tourism makes the most significant contribution to the economies of Spain and Greece and is less important in the larger industrialized economies of Germany and the United Kingdom. Germany's position as the major source of international tourists is confirmed by the largest absolute and relative expenditure on travel of all the member states.

These figures, however, say little about the net contribution of tourism to the balance of payments position of any particular country. A common though somewhat misleading attempt to rectify this is to present the net balance of tourism which is derived by subtracting expenditure by outgoing nationals from revenue brought in by foreign visitors. Thus in 1984 Spain would have had a net positive balance of 8801 million ECU while Germany would have recorded a deficit of 10,733 million ECU. This is largely spurious reasoning in that the two sums are fairly independent of each other (Gray, 1970). The net balance of payments contribution from international tourism can be calculated more accurately by relating expenditure by foreign tourists to associated costs. These costs and leakages may take several forms (IUOTO, 1975):

1. The costs of imported goods and services used by tourists; for example, imported fruit or whiskey.
2. The foreign exchange costs of capital investment in tourist facilities.
3. Payments abroad in the form of:
   (a) profits and capital remittances by foreign tourist companies,
   (b) wage remittances by expatriate workers,
   (c) interest payments on foreign loans,
   (d) management, royalties and other fees, e.g. for franchised hotels,
   (e) payments due to foreign travel agents and tour operators.
4. Promotion and publicity abroad.
5. Overseas training of personnel.
6. Extra expenditure on imports resulting from consumption by residents who have earned income from the tourist industry or whose consumption patterns have altered due to the demonstration effects of tourism.

Calculating these leakages and the net foreign exchange earnings of tourism is a difficult though not impossible task as recent studies in the Caribbean (Seward and Spinard, 1982), Asia (Pye and Lin, 1983) and the South Pacific (Britton, 1987) indicate. Table 6.3 depicts the elements of this exercise for Sri Lanka where import leakages accounted for about a quarter of gross earnings from international tourism in 1979.

Table 6.2 International tourism and balance of payments in the European Community (1984)

Member state	Revenue (mn ECU*) %†		Expenditure (mn ECU*) (%†)	
FR Germany	6790	2.6	17,692	7.2
Belgium/Luxembourg	2115	2.1	2484	2.5
Denmark	1639	5.8	1559	5.1
Greece	1678	18.0	431	3.1
France	9644	4.6	5413	2.6
Ireland	609	4.3	522	3.2
Italy	10,931	8.7	2669	2.0
Netherlands	1948	1.8	3834	3.9
UK	7046	3.9	7807	4.4
Spain	9868	20.8	1067	2.1
Portugal	1211	13.3	284	2.3

(*Source*: after Lee, 1987).
* ECU: European Currency Unit. 1 ECU = £0.60 (1984).
† % of current balance totals broadly defined.

Table 6.3   Foreign exchange costs, import leakage and net foreign exchange
receipts in Sri Lanka (1979)

Items	RS (000)
Current operating costs – registered establishments	168,733
Foreign debt servicing	12,477
Foreigh exchange costs of capital investment	34,128
Tourist Board expenditure	33,690
Tourist consumption outside the registered establishments	34,867
Allowance for the use of general infrastructure and public utilities (notional)	24,100
Allowance for the leakages through other miscellaneous sources (notional)	12,050
Total import leakage	320,045
Gross foreign exchange earnings from tourism	1,205,000
Net foreign earnings from tourism (foreign exchange surplus)	884,955
Import leakage as a percentage of gross earnings	26.6%
Foreign exchange surplus as a percentage of gross earnings	73.4%

(*Source*: Attanayake *et al.*, 1983).

Leakages from other countries reported in these studies are depicted
in Table 6.4 which shows that up to half of gross foreign exchange
earnings from tourism in small Caribbean and Pacific states may be
lost to the national economy. Comparatively little attention appears
to have been paid to estimating such leakages in developed countries
where the percentages are generally assumed to be much lower.

Table 6.4   Leakages of gross foreign exchange earnings from tourism
in selected Pacific, Caribbean and Asian States

State	year	% leakage of gross tourism receipts
Fiji	1979	56
Cook Islands	1979	50
St Lucia	1978	44.8
Aruba	1980	41.4
US Virgin Islands	1979	35.9
Antigua	1978	25.2
Hong Kong	1973	41
Sri Lanka	1979	26.6
Philippines	1978	10.8
Korea	1978	19.7

(*Source:* Britton, 1987; Seward and Spinard, 1982; Pye and Lin, 1983).

A New Zealand study indicated the net foreign exchange earnings of foreign tourism to be 88 per cent in 1976–77 (BERL, 1982). The import content of tourism in Yugoslavia was estimated to be only about 11 per cent of foreign exchange earnings in 1972 (Allcock, 1986). Allcock also points out that the prices obtained for goods sold to tourists within the country, for example the 14,000 tons of fresh meat and fish consumed by foreign tourists in 1979, are much greater than had they been shipped abroad in the form of conventional exports.

The extent of leakages and the degree to which goods and services consumed by foreign tourists can be provided domestically can vary significantly and will depend on (IUOTO, 1975; Theuns, 1976; Seward and Spinard, 1982; Pye and Lin, 1983; Britton, 1987):

1. The size of the nation.
2. The structure and diversity of the national economy.
3. The nation's import policy.
4. Whether or not supply can keep pace with demand.
5. The type of tourism and process of development.
6. The class of visitor.
7. The location of development – remote areas may draw more on imported goods due to the uncertainty of domestic supplies.

The focus on developing countries may also explain why the effects of these leakages on the beneficiaries, notably the developed countries generating the tourists and providing many of the inputs, appear to have been overlooked. Baretje (1982), however, proposes a broader system of accounting, the tourism external account (TEA), which would enable many of these inflows to be recorded (Table 6.5). In reporting on first attempts to estimate the TEA for France and Tunisia, Baretje notes in the case of France 'spectacularly different' results between the TEA and more conventional calculations of the 'balance of tourism', with a much greater positive balance being recorded. An 'ideal tourism balance' similar to the TEA is put forward by Sessa (1983). This broader approach clearly merits further attention but the difficulties of operationalizing it should not be underestimated.

As with other measures, tourism's performance in terms of its foreign exchange earnings capacity also needs to be assessed alongside that of other sectors. In the New Zealand study, tourism (88 per cent) was favourably placed alongside the much larger agricultural sector (87 per cent) and ahead of manufacturing (72 per cent), being surpassed only by other invisibles, that is non-tourism service exports (96 per cent) (BERL, 1982). Even though the leakage from tourism in Hong Kong was high compared to other Asian countries, it was still lower than the electronics industry, another important sector of the colony's economy. A similar situation was observed in Korea (Pye and Lin, 1983). Rao (1986) in an analysis of instability indexes for Fiji's foreign exchange

Table 6.5 Components in tourism's external account

Expenditures	Receipts
Expenditures by tourists abroad	Expenditures 'at home' by foreign tourists
Transportation	Transportation
Investments (outward)	Investments (inward)
Dividends, interest and profits paid out	Dividends, interest and profits received
Commodity Imports (tourism induced)	Commodity exports (tourism induced)
Capital goods	Capital goods
Consumption goods	Consumption goods
Salaries repatriated	Salaries sent abroad
Training	Training
Publicity and promotion	Publicity and promotion
Miscellaneous services	Miscellaneous services

(*Source*: Baretje, 1982).

earnings from tourism, total commodity exports and sugar, found that gross earnings from tourism were the most stable source of foreign exchange in the country over the period 1963–81.

Employment

As a service industry it is generally argued that tourism is labour-intensive and that one of the major impacts of tourist development, particularly at a regional and local level, is job creation. In 1980 employment attributable to tourist spending in the United Kingdom was estimated at about 1.2 million (BTA, 1982). This represented 4.9 per cent of total employment whereas tourist spending accounted for only 3.8 per cent of GDP, an indication of the labour-intensiveness of the tourist industry and perhaps also its lower wage rates. Labour demands may vary significantly from one sector of the industry to another and within sectors. A comprehensive economic survey of the Greater Tayside region of Scotland (Henderson, 1975) showed that for every 10,000 tourist days spent in the region, thirteen job opportunities were created there. However, hotel users were found to be responsible for more than twice this average amount (twenty-eight regional jobs per 10,000 hotel user days) while those using self-catering accommodation generated an average of approximately nine jobs.

In Ch. 4, a trend towards increased self-catering in times of economic recession was noted. More generally, where labour costs are high, as in many resorts in Western Europe, both developers and tourists are increasingly favouring accommodation with a low labour content such as rental apartments and condominiums. There are also signs of increasing productivity both in the United Kingdom and elsewhere. Ruiz (1985) notes that in 1972, eighty-eight workers were needed to produce 1 million dollars of output in Puerto Rico's hotel sector; by 1979 this number had fallen to fifty-nine (this decrease might also reflect some change in service provided). In Sri Lanka, labour productivity in the tourist sector was found to be somewhat higher than in the industrial sector (Attanayake, Samaranayake and Ratnapala, 1983), though this may not be the case in other countries.

A further advantage seen with tourist development is that it may create jobs more cheaply than other sectors of the economy. Widely varying results are reported in the literature, however. The BTA study (1982, p. 15) concluded 'in broad terms it would appear that the capital required per employee in tourist-related activities is about half the national average'. Capital employed per worker was about £9000 in hotels and catering in 1980 compared with an average of £19,000 in the economy as a whole. The capital–employment ratio for the Sri Lankan tourist industry 'compares well with most of the modern manufacturing industries established either to earn or save foreign exchange' (Attanayake *et al.*, 1983, p. 334). Romsa (1981), on the other hand, reports on a West German study which indicated the amount required to create a permanent job in tourism was about five times as large as in industry. Conflicting evidence regarding the capital–employment ratio advantages of tourism relative to other sectors is reported for the early stages of tourist development in Mexico (Mathieson and Wall, 1982). Significant differences may also occur between tourism sectors, with accommodation often being the most capital–intensive. Moreover differences within these sectors are also found, large luxury hotels will require heavier capital investments and may generate more jobs per room, but their capital–employment ratio may well be less than that of smaller, more modest hotels (Elkan, 1975).

Hughes (1982, p. 173), however, points out that comparisons with employment generation in other sectors 'are limited by the choice of opportunities for job creation in particular regions. Cost per job does not imply that there is a choice of job type that can be created, only that those that have been created have, perhaps fortunately, been cheaper.' To a certain extent, this argument could also be applied to the type of tourism developed.

Questions arise, however, of whether jobs in tourism are 'real jobs', with critics pointing out that much of the demand is seasonal and for female and part-time employees and that much tourist industry employment is characterized by low status and low pay (Vaughan and Long, 1982). In Greece about half of those employed in tourism work for only part of the year (Papadopoulus and Mirza, 1985), with significant variations occurring from sector to sector (Spartidis, 1976). In Britain, women account for more than 60 per cent of the hotel and accommodation work-force, with half the female staff being part-time (BTA, 1982). Wages in the hotel, catering and retail distribution sectors in Britain and tourism-related activities in Australia and West Germany are reported to be significantly lower than those in other sectors (Vaughan and Long, 1982; Cooper and Pigram, 1984; Romsa, 1981). Allcock (1986) details aspects of the tourist industry in Yugoslavia which he sees (p. 586) as having 'a poorly paid workforce skewed in its composition towards the lower end of the skill and educational scales, characterized by high levels of seasonality'. However, average salaries of hotel employees in Greece are estimated to be higher than those in other sectors (Spartidis, 1976), while in Sri Lanka the average monthly earing of tourist industry employees is much greater than in the industrial sector (Attanayake *et al.*, 1983). The variations noted here may reflect differences in the broader economic structure of developed and developing countries as well as in the patterns of tourist development experienced.

The degree to which the characteristics just outlined effectively reduce tourism's contribution to employment generation will depend on several factors. The effect of seasonality will be influenced in large part by the existence or not of other seasonal labour demands or opportunities and the extent to which these are complementary or competitive (Lever, 1987). Ski-fields, for example, require labour at a time when there is relatively little agricultural activity in alpine communities (Reffay, 1985). The demand for part-time female labour may be compatible with other domestic responsibilities. Much of the perceived impact of women working in tourism is conditioned by more general attitudes to female employment and whether or not this is considered appropriate. In Tunisia, women are largely absent in functions which are exercised in public places (restaurants, cafeterias, bars) and female participation is lowest in the most backward areas where cultural traditions are strongest (van Houts, 1983). In Scotland, Duffield and Long (1981, p. 414) conclude:

> The generation of female employment is to be welcomed, especially in those rural areas where traditionally economic activity rates among women have been low, but clearly tourism does little to alleviate the key structural problems of continuing male unemployment in the depressed rural economy. The lack of employment for primary wage earners (usually male) is a critical factor in precipitating out migration.

Above all, however, tourism may be able to provide jobs where little other potential exists and retain these jobs in changing economic conditions. As Getz (1986a, pp. 124–5) comments with regard to the Spey Valley in Scotland: 'Jobs created by a single manufacturing industry might very well be suited to attracting and holding families, but tourism is less susceptible to total collapse. When a one-industry community faces a closure, its entire economic base can be detroyed.'

Regional development

Regional development was one of the major aspects of the theme of development as distributive justice which emerged during the 1960s (see Ch. 1). While the goals of regional development may vary, the prime concern is usually 'to even out, or at least narrow the gap in the life chances, employment opportunities and real income of citizens irrespective of the region of the country in which they live' (Mabogunje, 1980). The purpose of the European Regional Development Fund (ERDF), for instance, is: 'to contribute to the correction of the principal regional imbalances within the community by participating in the development and structural adjustment of regions whose development is lagging behind and in the conversion of declining industrial regions' (*Official Journal of the European Communities*, 28 June 1984).

Tourism may constitute a useful tool for regional development for a variety of reasons. Firstly, the analysis of demand in Ch. 4 indicated that large metropolitan areas were major tourist markets, with the propensity for holidaytaking being highest there. Secondly, although metropolitan areas may constitute significant destinations in themselves, particularly for wanderlust tourists, coastal, rural and alpine regions are also much favoured destinations (Ch. 3). As these latter regions are often those which fall below the national average on indicators of socio-economic well-being, it is argued tourism can serve as a means of redistributing wealth geographically from the richer metropolitan areas to the poorer, peripheral regions. Moreover, the resources which attract tourists (Ch. 5) may not lend themselves readily to exploitation for other purposes such as manufacturing or agriculture. Duffield and Long (1981, p. 409), for example, note in the case of the Highlands and Islands of Scotland: 'Ironically, the very consequences of lack of development, the unspoilt character of the landscape and distinctive local cultures, become positive resources as far as tourism is concerned.'

For these reasons, certain regional development agencies have included tourism as a major element in their strategies at a comparatively early stage, though these regional goals have also often been set alongside other national ones. Examples here include the major coastal projects in Languedoc–Roussillon and Aquitaine co-ordinated by France's DATAR, the activities of the Highlands and Islands Development Board (HIDB) in Scotland (Duffield and Long, 1981), tourist development projects in Eastern Europe (Tassin, 1984), attempts to develop tourism on outer islands (Ch. 3) and a growing recognition of tourism potential through grants of the ERDF (Pearce, 1988a). In the first 10 years of the ERDF's existence (1975–84), grants totalling 166 million ECU (European Currency Unit, as at 1 December 1984, 1 ECU = £0.60) were made to tourism projects in member countries with the United Kingdom (60.8 per cent) and Italy (19 per cent) attracting the largest shares of all the member States. This accounted for only 1 per cent of all ERDF assistance during this period but significant increases in tourism grants were made in 1985 after tourism had been specifically identified in the new 1984 ERDF regulation as one of the sectors capable of exploiting the potential for, 'internally generated development'. In other cases, more spontaneous processes of tourist development or projects undertaken to meet other goals, for example foreign exchange generation, may also have an effect on regional development, either alleviating or accentuating regional economic imbalances (Odouard, 1973; Peppelenbosch and Templeman, 1973).

Tourism's capacity to reduce regional imbalances and contribute to 'internally generated development' will be influenced by factors similar to those affecting the extent of foreign exchange earnings. Leakages occur at a regional as well as a national scale, the degree of these being determined in large part by the nature of tourist development. Where most of the input required can be provided from within the region and development is largely within the hands of regional entrepreneurs and residents, then many of the benefits which arise from the development will remain within the region. Conversely, where tourist development depends primarily on external inputs of capital, labour, know-how and technical resources, then leakages from regional economies can be very high. Such is the case in areas where the absence of adequate infrastructure contributes to enclave development (Ch. 3).

Duffield and Long (1981, p. 415) provide a very useful review of tourism's contribution to regional development in the Highlands and Islands of Scotland:

> Whatever the weakness of tourism-related employment [e.g. a high proportion of female workers and non-resident employees], it has been estimated that some 30,000 people are employed within the HIDB area in jobs directly or indirectly connected with the tourist industry, and that therefore as much as 30 per cent of the working population owes its employment in one way or another to tourist activity. The HIDB . . . has estimated that tourism now earns approximately one quarter of the Gross Highland Product.

In drawing attention to the distribution of these impacts, Duffield and Long note that the HIDB investment has favoured those areas which historically have attracted the greatest tourist numbers. But as they also point out (p. 417):

In the Highlands and Islands the tendency to create prosperous enclaves of tourist activity within rural areas is however mitigated against to some degree by the nature of tourist activity itself. Although certain centres are particularly popular (with half the visitor nights being spent in 20 centres) a distinctive feature of tourism in the Highlands and Islands is the touring holiday, during which visitors stay in a number of locations. In 1973 half of all holiday visits were touring holidays. This tendency, coupled with the fact that three quarters of visitor groups travel by car, means that holiday visitors penetrate deeply into the rural areas . . . thus ensuring the widespread distribution of both economic and social impacts upon rural communities.

In India, the redistributive effects of tourism are less apparent (Pavaskar, 1982, p. 36):

> Evidently, despite the development of many tourist resorts throughout the country since the advent of planning in 1951, not only do the four metropolitan centres continue to attract more tourists, both foreign and domestic, but they also seem to encourage them to stay longer and spend more. Apparently, even though a few backward areas have benefited from the development of tourism, the continued flow of tourist money to the large metropolitan cities has not much helped to reduce the widening regional disparities in economic development. The rapid rise in the number of approved hotels and posh restaurants in all the metropolitan cities in recent years bears an eloquent testimony of such failure.

A more quantitative and positive analysis of tourism's role in redressing regional imbalances is provided by Allcock (1986) with respect to Croatia in Yugoslavia. From the trends in Table 6.6, Allcock concludes (p. 573): 'Although few of these settlements feature among the most affluent of the municipalities in the republic (and there have been some backward movements) on the whole tourism does seem to have been associated with a quite remarkable improvement in the economic situation of areas which were previously rather poor.'

Dorfmann (1983, p. 20) is more cautious about measures of regional economic development and suggests in his review of development policies in the European Alps that:

> these policies may have permitted mountain regions to experience a certain quantitative growth (an improvement in the demographic situation and a per capita increase in revenue) but they have not triggered off a veritable process of economic, social and cultural development. In other words, the policies for mountain regions have prevented them from falling into the vicious circle of poverty but they have not generated autogenous growth.

This situation, particularly as it involves tourism, is characterized by sectoral and regional imbalances as certain resources and areas are exploited to meet non-local demands and by a dependence on external, primarily urban, inputs (capital, assistance, plans, etc.). Dorfmann advo-

Table 6.6 Tourism and prosperity: Croatia's leading tourist communes

Commune	Rank ordering by number of tourists	Rank ordering by over- night stays of foreign tourists	Rank ordering by per capita national income 1961	Rank ordering by per capita national income 1981	Change in rank ordering + or −
Dubrovnik	1	2	24	15	+9
Porec	2	1	61=	6	+55
Makarska	3	7	49=	24	+25
Buje	4	3	34	5	+29
Crikvenica	5	11	48	31	+17
Opatija	6	6	64=	25	+39
Sibenik	7	8	26	41	−15
Pula	8	5	19	23	−4
Zadar	9	12	36	42	−6
Rovinj	10	4	8	14	−6
$\bar{x}$			36.9	22.6	+21.25

(*Source*: Allcock, 1986).

cates a new form of 'integrated' development, one which incorporates local residents to a greater degree and is aimed more at meeting their needs than external demands. It is not clear, however, how and where alternative forms of tourism fit into Dorfmann's proposed strategy.

In Corsica, non-acceptance by local residents of large-scale coastal tourism projects characterized by increasing dependence on imports and outside capital led to the disbandment of the government's regional development agency charged with promoting tourism, SETCO (Societé d'Équipement Touristique pour la Corse), in 1977, with very few of the projected developments being completed (Kofman, 1985; Richez, 1986). Nevertheless, tourism has expanded significantly on the island with many of the large coastal complexes being externally controlled.

The multiplier effect

Much discussion of tourism's contribution to regional development, and to economic development in general, concerns the way in which expenditure on tourism filters throughout the economy, stimulating other sectors as it does so. This is known as the 'multiplier effect'. The multiplier is a measure of the impact of extra expenditure introduced into an economy, in this case through tourism. In most studies account appears to be taken solely of tourist spending though provision might also be made for expenditure in the form of investment by external sources in tourism plant and infrastructure, and the export of goods stimulated by tourism (WTO/Horwath and Horwath, 1981).

Tourism multipliers reflect the interrelationships of three types of expenditures (WTO/Howarth and Horwath, 1981; Archer, 1982):

1. Direct expenditure – that resulting from tourist spending on the full range of goods and services sought by them, including tourism-generated exports, or from tourism-related investment. Account is taken only of that expenditure which initially remains in the area and is not lost through the leakages discussed earlier and through savings.
2. Indirect expenditure – money remaining in the area may be re-spent locally in successive rounds of business transactions, as, for example, hoteliers purchase goods from local retailers or in turn acquire them from wholesalers. As a consequence of these transactions, the general output of the area rises, employment opportunities increase and personal incomes will rise. At each successive round, further leakages will occur until little or no further re-spending is possible.
3. Induced expenditure – additional personal income will generate further consumer spending, for example as hotel workers purchase goods and services with their wages.

As Archer (1982, p. 237) notes: 'The indirect and induced effects together are sometimes called *secondary* effects, and the tourism multiplier is a measure of the total effects (direct plus secondary) which result, from the additional tourist expenditure.'

Initially some impressive claims concerning the multiplier effect of tourism were put forward, in the Pacific (Clement, 1961) and the Caribbean (Zinder, 1969). These no doubt contributed to the early optimistic view of tourism's role in economic development discussed in Ch. 1 but were later considered exaggerated by other writers who advanced much lower multipliers (Bryden, 1973). Much of the early confusion in the literature arises out of a failure to recognize that different types of multipliers exist, that different methodologies have been used, that scale is important and that different economic contexts will give different results.

Useful reviews of the four different types of multipliers in common use are provided by Archer (1977, 1982), Hanna (1976), WTO/Horwath and Horwath (1981), and McCann (1983). These are:

1. Sales or transaction multipliers measure the extra business created (direct and secondary) by an extra unit of tourist expenditure.
2. Output multipliers are similar to (1), but in addition to sales take account of changes in stocks.

3. Income multipliers show the relationship between an additional unit of spending and the changes in the level of income in the economy. They can be expressed in either of two ways. The ratio method, which expresses the direct and indirect incomes (Type 1) or direct and secondary incomes (Type 2) generated per unit of direct income while the normal method expresses total income (direct and secondary) generated in the study area per unit increase in final demand created within a particular sector. The normal method is considered to be the most useful in these reviews while ratio multipliers indicate the internal linkages which exist between various sectors of the economy 'they give no indication *per se* of how much tourism expenditure is needed to create a unit of direct income' (Archer, 1982, p. 238).

4. Employment multipliers are expressed either as the ratio of the direct and secondary employment generated by additional tourism expenditure to direct employment alone or as the amount of employment created by tourism per unit of additional expenditure.

These different types of multiplier are intrinsically linked as Archer (1982, p. 238) points out:

... additional tourist spending of, say 1 million pound may generate 2.5 million pound of output within an economy and 0.5 million pound of direct and secondary income to nationals in the area. It may create also 200 extra jobs and perhaps 180 secondary jobs. The output multiplier, therefore, is 2.5, the income multiplier 0.5 and the employment multiplier 1.9 ($^{380}/_{200}$) or 3.8 (i.e. 3.8 jobs) per 10,000 pound of tourist spending.

Two methods of deriving tourism multipliers are used: *ad hoc* models and input/output analysis. The *ad hoc* model is an adaptation of the classical Keynesian multiplier in the form:

$$K = A \times \frac{1}{1-BC}$$

A = Per cent of tourist spending remaining in the region after first round leakages.

B = Per cent of income of local residents spent on local goods and services.

C = Per cent of expenditure of local residents that accrues as local income, i.e. minus other leakages.

Thus, if 50 per cent of tourist spending remains after first-round leakages and if residents spend 60 per cent of their incomes locally and

40 per cent of that accrues as local income then the income multiplier
(*K*) is:

$$0.5 \times \frac{1}{1 - 0.6 \times 0.4} \text{ or } 0.65$$

The *ad hoc* multiplier can also be further refined by disaggregating
tourism into component parts. Calculation of the multiplier in this way
will usually require the generation of substantial survey data to provide
the necessary inputs into the model.

Where regional or national input/output tables exist, the second
technique can make use of available data (computation of such tables
in the first place is also expensive and time-consuming). As Archer
(1982, p. 238) notes:

> Input–output analysis is a method of tabulating an economic system in
> matrix form to show as rows the sales made by each sector of the economy
> to each of the other sectors and as columns the purchases made by each
> sector from each of the others. Tourist spending is shown as an export
> column and, by means of matrix algebra, the impact of this expended on
> each sector and on incomes can be measured.

For a more technical discussion of these methods and a review
of the weaknesses and limitations of tourism multipliers, see Archer
(1973, 1977, 1982), Hanna (1976), Pigram and Cooper (1980),
WTO/Horwath and Horwath (1981) and Hughes (1982). Most
multiplier studies, for example, make assumptions that an increase in
output required to meet increased tourist demand can be met from the
same sources (i.e. supply is elastic) and that no changes will occur in
patterns of transactions as expenditure increases or in consumption as
income rises.

Table 6.7 depicts tourist income multipliers compiled by Archer
(1982) from a range of studies undertaken in the 1960s and early
1970s. These show that the value of the multiplier depends on the
size and nature of the economies concerned, and on the degree of
sectoral linkages within them. While values vary from place to place,
there is a certain amount of consistency in like areas, notably in the
set of British counties and regions and in the group of UK towns and
villages. According to Archer, the small islands recording values near
or above unity either have well developed linkages within the economy
to either an agricultural or manufacturing base (e.g. Dominica and Hong
Kong) or have a high value added on the goods or services provided to
tourists (e.g. Bermuda and the Bahamas). Lower values are reported for
islands with few sectoral linkages and high leakages (e.g. Cayman and

Table 6.7   Tourism income multipliers in selected areas

	Tourism income multipliers
(a) Small island economies	
Dominica	1.20
Bermuda	1.03
Hong Kong	1.02
Indian Ocean islands	0.95–1.03
Hawaii	0.90–1.30
Antigua	0.88
Bahamas	0.78
Fiji	0.69
Cayman Islands	0.65
British Virgin Islands	0.58
(b) US States and counties	
Hawaii	0.90–1.30
Missouri	0.88
Walworth County, Wisconsin	0.77
Grand County, Colorado	0.60
Door County, Wisconsin	0.55
Sullivan County, PA	0.44
Southwestern Wyoming	0.39–0.53
(c) UK counties and regions	
Gwynedd, North Wales	0.37
Cumbria	0.35–0.44
South West England	0.33–0.47
Greater Tayside	0.32
East Anglian coast	0.32
Lothian Region	0.29
Isle of Skye	0.25–0.41
(d) UK towns and villages	
Carlisle	0.40
Great Yarmouth	0.33
Kendal	0.28
Appleby	0.25
Sedbergh	0.25
Keswick	0.24
Towns and villages in Wales	0.18–0.47

(*Source*: Archer, 1982).

British Virgin Islands). Archer attributes the lower multiplier values of
the UK areas compared to the US States and counties to the extent to
which they trade with the rest of the country.

Var and Quayson (1985) make a similar comparison between the
Okanagan region of British Columbia (income multiplier value of

0.731) and the British regions and attribute the difference to 'the greater degree of interdependence and diversification of the Okanagan economy relative to the regional economies of the UK'. Conversely, the employment multipliers in the UK regions are much larger than in the Okanagan and in Victoria BC. Var and Quayson see this as a function of differing patterns of tourist expenditure, notably to proportionately more spending in the retail sector in the British regions and spending on recreation in the Canadian ones.

Different multiplier values may also be found within regions between different forms of tourism and types of tourist. The British studies indicate that bed-and-breakfast places usually produce greater income multipliers because more of the expenditure is retained locally than with other types of accommodation (Archer, 1973; Hughes, 1982). From their analyses of different types of accommodation in Victoria, Liu and Var (1982, p. 187) concluded that 'smaller multipliers were generally found for the central, the large, the affiliated and the externally owned establishments because of smaller linkages in the local economy'. A similar pattern occurs in the Cook Islands where the income and employment multiplier coefficients were found by Milne (1987, p. 59) to be 'inversely related to increasing levels of overseas control, organisational complexity, and firm size'.

In the Cook Islands motel guests generated more income per dollar spent (48 cents, cf. 39 cents) and more employment (0.94 jobs per $10,000, cf. 0.74) than hotel guests. Nevertheless, as a consequence of their higher daily expenditure, hotel guests each generated nearly double the amount of income and employment created by motel guests. Multiplier coefficients also varied by package type and to a lesser extent by nationality. Within the Okanagan, convention delegates produced the highest income multipliers whereas non-residential and residential (BC) tourists had larger employment multipliers (Var and Quayson, 1985). Foreign and domestic tourism in India do not generate significantly different output and income multipliers but the latter has a higher ratio employment multiplier (3.58, cf. 2.82 for foreign tourism) due to the less capital-intensive forms of accommodation used by Indian tourists (Pavaskar, 1982). Similarly in Turkey, Liu, Var and Timur (1984) found only negligible differences to total tourist income multipliers for various groups of foreign and domestic visitors, although some variations occurred on the direct and induced components of the multiplier for each group.

Analysis of different types of tourism multipliers and the values of different forms of tourism must not, however, obscure the fact that the multiplier effect is not confined solely to tourism as some studies seem to imply. How does tourism compare in this regard to other sectors? In a frank review of multipliers in Britain, Hughes (1982, p. 172) concluded: 'Tourism multipliers would appear to perform averagely well in comparison with other regional multipliers. They are not consistently superior and do not warrant the special status accorded to them.' In a better documented study at the national level in Australia, Cooper and Pigram (1984, p. 12) found:

> . . . tourism has quite weak linkages with other sectors of the economy and therefore may have little impact as a propulsive industry for economic growth. It has a characteristically labour-intensive employment structure and a high propensity to be concerned with rises in personal income but has a relatively slight multiplier effect on output and income, compared with other sectors of the economy.

Inter-sectoral competition

Although tourism may stimulate other sectors of the economy, it may also disrupt or compete with them, especially where there is a shortage of labour or investment capital. That is to say, there are opportunity costs associated with the development of tourism, as there are with any new economic activity – resources devoted to tourist development cannot be used in other sectors. The extent of these opportunity costs will depend in part on other development possibilities. Duffield (1982, p. 250) notes in the case of Scotland that ' . . . many areas lack the luxury of diversification and, given the desire that people should stay and work, for example, in remote rural areas, in the short term at least, the opportunity cost of investment in tourism is low'.

Some of these issues can be illustrated well with regard to the relationships between tourism and agriculture. A particularly striking case is that given by Long (1978), of the development of tourism in the British Virgin Islands. This led to a massive increase in GNP but the impact on agriculture was less satisfactory. The demand for agricultural imports increased from $1.8 million in 1969 to $4.3 million in 1974 and there was a significant decline in national agricultural output. This was attributed to a redeployment of the bulk of farm labour in the tourism sector, the displacement of local products by competition from North American ones and banks giving credit to businessmen rather than farmers. However, Latimer (1985), in a comprehensive review of tourism versus agriculture in developing island economies over the peak years of the 1960s and early 1970s, challenges the widely held view that tourism was destroying agriculture and other sectors by competing for labour and land and by increasing their reserve price. Latimer firstly notes that much of the evidence put forward by Bryden (1973) and others was related to small tropical islands with limited growing potential, many of which were already suffering from the long-term decline in sugar and copra trades, and that a different situation may prevail in larger developing economies such as Kenya and Tunisia. But even in the small Caribbean islands he argues (p. 41):

> it is questionable whether tourism took sizeable resources of land and labour away from agriculture and even whether there was any net decline in agriculture at all, though more facts are needed on this issue. It has been claimed that better use for public money could have been found in agriculture than in supporting the infrastructure of tourism but this seems unlikely. It is agreed that more could have been done to aid the development of agricultural and fishing inputs into tourism and other growth-related demand.

A similar situation is found in the Mediterranean where the literature in the early 1970s stressed agricultural labour shortages and enormous rises in land prices as two common consequences of tourist development. According to Hermans (1981, p. 465): 'The influence of tourist development on agriculture it seems, is generally perceived as problematic at best, disastrous at worst.' However, she suggests, these judgements were based on very little evidence and in a well documented case study of Cambrils on the Spanish Costa Dorada shows tourism promoted agriculture directly and indirectly, for example by allowing diversification of markets through improved access. Hermans also acknowledges the importance of local factors and while she is wary of making generalizations, points out that the encounter of agriculture and tourism 'is not always deleterious'. Elsewhere in Spain, Boaglio (1973) suggests that by creating a demand for labour and stimulating rural depopulation, tourism may effectively lead to a modernization of agricultural production.

Inflation

The development of tourism may also have a general inflationary effect, especially during the initial stages when the supply of goods and services can often not respond quickly enough to meet the increased demand. Moreover, there is often a significant disparity between the spending power of the tourists and the host population. Housing prices in particular may rise very quickly, with tourists seeking holiday homes and external developers and employees looking for rental or permanent accommodation. Seasonal fluctuations in the price of food are often exacerbated by tourism. These effects have not yet been well documented and difficulties arise in separating out tourism-induced effects in periods of more general inflation. Higher costs of living were identified in an attitudinal survey of residents as one of the major disadvantages of the expansion of tourism in Queenstown, New Zealand, with some 80 per cent of respondents feeling they were worse off (Pearce and Cant, 1981). A comparison of municipalities in the Graubunden canton of Switzerland showed land prices in those with a large proportion of second homes increased by 43 per cent over the period 1967–73, compared to 8 per cent in those which had none (Federal Department of Forestry, 1981).

State Revenues

The main State benefits to be derived from tourist development will result from greater tax revenues (incomes, corporate, sales, property), increased overseas earnings, reduced social charges (e.g. unemployment benefits) and perhaps profits from direct intervention in the industry (e.g. State hotel chains). A very useful review of tourism's contribution to State revenues is given in a WTO (1983c) report which outlines the different fiscal regimes of a range of developed and developing countries and variations in tax revenues generated by tourism. Table 6.8 shows that in small States heavily dependent on tourism over half of the total tax revenue may come from tourism while in larger, industrialized countries this proportion is commonly

Table 6.8 Contribution of tourism to State revenues in selected countries

Country	Year	Fiscal revenue from tourism US dollars (millions)	Fiscal revenues from tourism	
			Total State revenues (%)	International tourism receipts (%)
Australia	1973/74	237.1	0.2	12.6
Bahamas	1968	—	55.0	—
	1974	74.6	61.9	22.7
Bermuda	1975	35.6	57.0	21.3
Canada	1974	380.9	1.1	25.6
Jamaica	1975	10.3	1.5	18.4
Kenya	1976	9.1	1.4	8.3
Pakistan	1978/79	21.9	1.0	21.9
	1979/80	23.9	—	15.6
Peru	1979	26.3	0.7	8.1
Seychelles	1976	0.6	—	5.6
Spain	1974	801.6	7.6	24.5
Sri Lanka	1971	4.9	1.0	15.4
United States	1977	—	3.4	—
of America	1978	13,985.0	3.4	—
Yugoslavia	1972	78.4	2.3	13.5
Zambia	1980	0.04	—	0.2

(*Source*: WTO, 1983c).

less than 5 per cent. Variations also occur in the ratio of tax revenues to international tourism receipts, with values of from 10 to 25 per cent being recorded in a number of countries. In countries such as the USA, much of the tax will come from domestic tourism with each domestic travel dollar spent there estimated to generate 30 cents in taxes (Mill and Morrison, 1985). A detailed study of Galveston, Texas, put the total tax take at between $11 million and $14 million annually, with more than a million dollars being retained by the city itself (Rose, 1981). While central government often derives much of its revenue from income and corporate taxation, regional and local governments frequently depend more on indirect taxes. Property taxes are especially important for local authorities; expansion of this tax base was one of the reasons noted in Ch. 3 for public sector encouragement of second homes.

Revenue from tourism will be reduced by the charges incurred in developing the industry; such as investment in infrastructure, development incentives, and promotional and training expenses. The infrastructure charges may be especially onerous for local authorities as this sector should be developed at the outset, requires reasonably large amounts of capital, brings little or no direct income and indirect returns in the form of taxes or rates are largely in the long term. New and heavy operating expenses may also be borne by the local authorities, for example, rubbish disposal, road maintenance, and snow clearance in ski-resorts. Bryden (1973) suggests that many of these public development costs were seriously underestimated in early evaluations of the impact of tourism in the Caribbean. A more recent study of overall costs and benefits in Sri Lanka indicated that 'every rupee spent by the Government on tourism both directly and indirectly resulted in a revenue yield of Rs 1.67 leaving a net benefit of 67 cents to the Government' (Attanayake *et al.*, 1983, p. 335). In a detailed cost/benefit study at the local level, Soesilo and Mings (1987) conclude that winter visitors to Scottsdale, Arizona, had a strong positive impact, contributing $2.5 million in direct sales tax to the city while creating a demand for twenty-one additional full-time public employees at a cost of $1.2 million, a revenue/cost ratio of 2.2 : 1.

Who benefits? Who pays?

Finally, in assessing these various costs and benefits, it is important throughout to identify just what groups or individuals are being affected. Four broad groups may be identified (Fig. 6.2). Firstly, there are those directly involved in the development process – the promoters, operators and their employees. Much of the direct return from tourism will accrue to this group. Then there are the other residents and enterprises who may not be directly engaged in any tourist activity but whose lives may nevertheless be affected by the expansion of tourism. Many of the indirect costs, such as the diversion of capital, land and labour and tourist-induced inflation, may be felt by this sector of the community, although benefits may be experienced as well, notably through the multiplier effect. The public authorities may be development agents, as has been noted in Chs. 2 and 3, but it is useful to distinguish them from the private sector in that the nature of their benefits and costs may differ significantly from those of the private sector. Finally, in paying for the various services they demand and use, the tourists bear many of the direct costs. Their benefits tend to be essentially non-monetary ones. Where, however, the advent of international tourism pushes up prices, extra charges may be incurred by the domestic holidaymaker. When the geographic origin of each of these four groups is considered along with their internal diversity, then very complex situations may exist and considerable variation in the overall pattern of costs and benefits occur.

Given the analytical difficulties associated with establishing various aspects of these costs and benefits, most studies are far from comprehensive and do not deal systematically with each of these four broad groups and much more work needs to be done in this field. As many of the studies cited in the preceding sections have indicated, emphasis has been given to determining local/non-local benefits and measuring the extent of leakages from local, regional and national economies. Milne (1987) gives a particularly useful and detailed account of the distribution of the economic impact of tourism in a small island economy, that of the Cook Islands. He notes (p. 44): 'Overseas operations clearly dominate the industry. They receive approximately 60 per cent of all tourist receipts, local Europeans receive 23 per cent, with the remaining 17 per cent flowing to Maori owned enterprises. Even if the Government's holdings in the industry are included, the Maori category only receive 38 per cent of total turnover, the same as overseas interests.'

Préau (1983) provides an excellent in-depth longitudinal study of family participation in the development of the Courchevel (France) ski-field and shows significant variation occurs in the ways and extent to which different families have been able to participate in and benefit from tourism. Préau bases his analysis on the changing modes of production in which a sample of forty families were engaged in 1948, primarily in peasant agriculture, and again in 1973. Passage to the capitalist mode of production as entrepreneurs was largely a function of the capacity to invest. This in turn basically depended on the size and location of existing land and property and by 1973 involved only nine of the forty families. The majority of the remainder, having fewer savings and capital, had also been drawn into the capitalist mode of production, but as paid labourers rather than business owners.

With the exception of some socio-economic surveys to be discussed later in the following section on social/cultural impacts, fewer studies have been concerned with the non-tourism oriented residents. One such study is that by Gray and Lowerson (1979) who observe in commenting on the 'two faces' of Brighton: 'Externally generated wealth remains of great importance, but much of it has been ploughed into projects whose value does little for the chronic problems of the town, particularly an inadequately housed low-wage labour force competing for resources with visitors and growing numbers of students.'

Nor has there been much research on the impact of tourism on non-destination areas. One notable exception is Lever's (1987) study of Spanish tourism migrants to Lloret de Mar. She concludes (p. 469) with regard to the source area of the migrants, a Castilian village:

> Tourism migration has brought few useful skills to the pueblo studied. There is not much demand for the migrants' linguistic skills for example, nor is there much productive investment as a result of remittances. Migration generally has been a way of shelving the problem of long-term development of rural areas, an issue which has become more urgent as opportunities for migrants have diminished.

ASSESSING THE SOCIAL AND CULTURAL IMPACT OF TOURIST DEVELOPMENT

Many studies of the social and cultural impact of tourist development were initially framed in terms of social and cultural costs but subsequent research has tended to be more balanced, acknowledging also the advantages and disadvantages which the expansion of tourism can bring to different societies and communities. A prime consideration in examining the social and cultural impact of tourist development is the nature and composition of the various groups involved and the relationships between these. The basic dichotomy of 'host and guests', popularized by the comprehensive anthropological volume of that name (V. Smith, 1977b) is generally accepted, though Jafari (1982) has proposed a tripartite cultural division. Jafari distinguishes between the broader imported culture associated with the guests and a more specific 'tourist culture' which (p. 57) 'refers to a way of life practised by tourists while travelling.' Tourists have been classified in different ways (Cohen, 1972, 1974; Plog, 1973) and variously segmented by market researchers (Ch. 4) but Smith's typology (Table 6.9) is particularly relevant for social/cultural impact studies. Demographic, social, ethnic and linguistic differences may also exist within the host population, certain sections of which may participate in or be affected by tourist development more than others. Fernandez Fuster (1974) also notes that the effects of tourism may be felt not only in the tourist centres but also

Table 6.9 Frequency of types of tourists and their adaptations to local norms.

Type of tourist	Numbers of tourists	Adaptations to local norms
Explorer	Very limited	Accepts fully
Élite	Rarely seen	Adapts fully
Off-beat	Uncommon but seen	Adapts well
Unusual	Occasional	Adapts somewhat
Incipient mass	Steady flow	Seeks Western amenities
Mass	Continuous influx	Expects Western amenities
Charter	Massive arrivals	Demands Western amenities

(*Source*: V. Smith, 1977b).

in neighbouring non-tourist towns and in the generating areas. Again, it is important to identify what particular groups or segments of society are experiencing what specific effects of tourist development. The social impact of tourism will vary according to the difference between the visitors and the visited, whether in terms of numbers, race, culture or social outlook. Lundgren (1972, p. 94), observes that in general: 'the force of tourist-generated local impact seems to increase with distance from the generating country'.

Several specific characteristics of tourism must also be kept in mind. Firstly, the transitory nature of the relationships between hosts and guests, often coupled with language barriers, allows little opportunity for understanding to develop between the two groups. Secondly, the fact that the tourist is on holiday while the host is at work may heighten differences between the two, especially as holiday behaviour is generally much less restrained than usual (Ch. 4). Thirdly, the seasonal nature of much tourism tends to be more disruptive than year-round activities, often creating the need for seasonal workers and exacerbating any tensions which might exist between the different groups. Finally, the outward signs of tourism may be more manifest than other types of development; for example, agricultural reform or the introduction of television. As a result, effects which have their origins elsewhere may be attributed to tourism. At worst, the industry may become the general scapegoat for any and every social *malaise*.

Table 6.10 summarizes many of the social and cultural impacts which tourist development may have. Firstly, tourism may affect the demographic structure of their population as studies have shown in such diverse contexts as the coast of Spain (Dumas, 1975b), the French Alps (Préau, 1983), the Highlands of Scotland (Getz, 1986b) and Nepal (Pawson, Stanford and Adams, 1984). Development of the industry will usually affect the size of the resident population, as the creation of

Table 6.10 The social and cultural impacts of tourism.

(a) Impact on population structure:
    size of population;
    age/sex composition;
    modification of family size;
    rural → urban transformation of population.
(b) Transformation of forms and types of occupation:
    impact on/of language and qualification levels;
    impact on occupational distribution by sector;
    demand for female labour;
    increase in seasonality of employment.
(c) Transformation of values:
    political;
    social;
    religious;
    moral.
(d) Influence on traditional way of life:
    on art, music and folklore;
    on habits and customs;
    on daily living.
(e) Modification of consumption patterns:
    qualitative alterations;
    quantitative alterations.
(f) Benefits to the tourist:
    relaxation, recuperation, recreation;
    change of environment;
    widening of horizons;
    social contact.

(*Source*: after Figuerola, 1976).

new jobs slows out-migration or attracts new workers or residents to the area. Ultimately there may even be a significant change in the degree of regional urbanization as development and migration tend not to be aspatial. These processes also tend to be age- and sex-selective, thereby altering the composition of the population as well as its size. Figure 6.3 depicts the age–sex pyramid of Queenstown, New Zealand, and shows that compared to the national average, the demographic structure of this major resort is characterized by an excess of residents aged from twenty to thirty, by a shortfall of under-fifteen-year-olds and by a larger proportion of females. Greater occupational and geographical mobility may also modify personal ties and family structures (Greenwood, 1972; Préau, 1983; Lever, 1987). In any given area, particular processes of development and local factors will combine to produce a set of population changes and impacts. Figure 6.4 shows tourism-related growth in the Badenoch–Strathspey district of Scotland has mainly

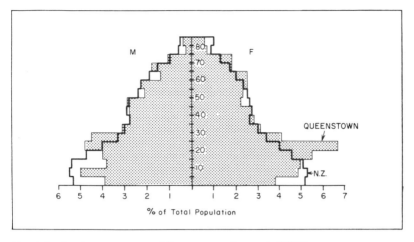

Fig. 6.3 Age/sex structure of Queenstown borough (1971).

been in Aviemore and other large villages, with smaller rural settlements continuing to experience population decline. Different structural changes and related social impacts have also occurred in the two sets of areas.

Occupational changes may also stem from tourist development. Language can be a critical factor in determining which sectors of the population will profit from tourism. Many native spectacles, for example, are presented by expatriate entrepreneurs or organized by the better-educated indigenes, the so-called 'culture brokers'. White (1974) has shown a more general decline in Romansch speaking in a multilingual area of Switzerland as tourism has developed there. The prospect of good jobs in the tourist industry may increase the desire for educational attainment. However, if initially the demand for qualified staff exceeds local capabilities or if control is in the hands of external promoters, then the better positions will be filled from outside, leaving the local residents the more menial tasks. Continuation of this policy will lead to frustration and perhaps hostility towards tourism. Employment opportunities in tourism may draw workers from other sectors of the economy – for example, agriculture – with consequent effects on class or social structure. Préau's (1983) detailed longitudinal study of the impact of Courchevel highlights the occupational changes which have occurred from one generation to the next. In his sample the percentage engaged in agriculture declined from 85 per cent in 1948 to 6 per cent in 1973, with nine jobs out of ten at this later date being

219

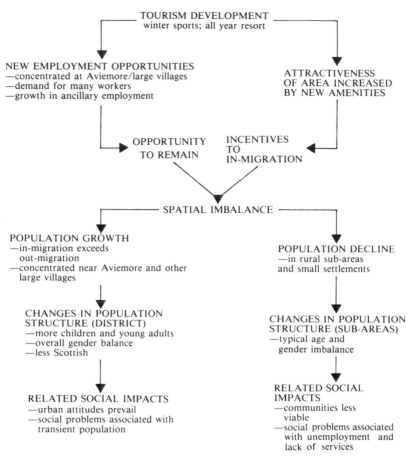

Fig. 6.4 Tourism-related population changes and their impacts in
Badenoch–Strathspey, Scotland.
(*Source:* Getz, 1986).

tourism-related. New occupational structures have also brought about attitudinal changes with further changes being anticipated amongst the third generation. Elsewhere, jobs which previously had no cash return in many societies, such as cooking and cleaning, may now become income-earning if performed in hotels, thus altering the status of these workers, particularly of women (van Houts, 1983; Cohen, 1984; Lever, 1987). Resorts dependent on significant influxes of seasonal workers may suffer from a lack of social stability, the constant turnover in the population allowing little time for lasting relationships or community spirit to develop. Added to this are the work patterns of the tourist industry, often characterized by long hours and split shifts. On the other hand, some, particularly the younger seasonal workers, may be attracted by the excitement and interest of continually meeting new people or welcome the freedom which leaving their more traditional settlements has brought (Lever, 1987).

Values may be transformed through the bringing together of different groups of people, even if at times only very briefly. This is commonly known as the 'demonstration effect'. On the Greek island of Myconos, for example, the sociologist Lambiri-Dimaki (1976) notes the: 'demo-cratization and modernization of attitudes' amongst the young, arising out of contact with youthful Western tourists. In this case it was the observed behaviour of the visitors, especially the equality of the females, which was as important as direct contact between the local residents and the tourists. Elsewhere, tourist behaviour may be offensive to local norms as with the arrival of hordes of scantily clad foreigners on the beaches of conservative and Catholic Spain.

Increased prostitution is commonly associated with an expansion of tourism, with vacations providing an opportunity for the inversion of sexual and other restrictions (Table 4.1). Sex tours to Korea, Thailand, the Philippines and other Asian countries have drawn strong criticism from religious, feminist and other writers (O'Grady, 1981; Graburn, 1983a; Fish, 1984). Jones (1978), in his brief study of prostitution in Bali, concluded that it is usually changed, not caused, by tourist demand. Cohen (1982), in a more comprehensive study of tourism-related prostitution in Bangkok, highlights the broader social and economic conditions which lead Thai women into prostitution, notes that 'the majority of prostitutes still serve a local Thai and Chinese clientele' and 'that the women working with *farangs* are in many respects the "élite" among the prostitutes'. Cohen emphasizes the complexity of the different relationships which may exist and concludes (p. 424): 'While it [prostitution] enables the girls to resolve some of the pressing problems caused by the precariousness of their life-situation, it does not usually resolve the problem of precariousness itself, but often exacerbates it.' Elsewhere, sex roles may be reversed, as with the 'beach-boy–Canadian secretary' syndrome of the Caribbean or the liaisons between young males and Scandinavian tourists in the Gambia (Wagner, 1979). Research in developed countries has been limited more to morphological studies, such as Burtenshaw, Bateman and Ashworth's (1981) examination of Amsterdam's red-light district.

In some cases it is the tourist who is influenced the most. This is especially the case with those seeking to reinforce or expand their political, ideological or religious beliefs by visits to such places as Israel, India, Cuba, Mecca and Lourdes. The oft-cited increase in worldwide understanding resulting from international tourism is perhaps exaggerated, given the nature and duration of tourist–host contact, but what information there is available suggests tourism has a positive effect in this domain. Survey research in the European Community, for instance, indicates attitudes towards the unification of Europe are stronger amongst those who have travelled, especially those visiting several countries, but the nature of this relationship requires further elaboration (Giourgas, 1985). Mings (1985, 1988) examined the attitudes towards and knowledge of the United States and American people amongst a sample of hotel employees and control groups of students in Barbados. Strongly positive attitudes towards American visitors were expressed by the hotel employees but the analysis of differences in levels of knowledge between them and the control groups proved inconclusive, in part due to the effect of information obtained through sources other than tourism. Mings (1988) concluded:

> The fundamental question remains: Does tourism help set the stage for improving international understanding? The results of this research suggest a qualified yes. However, perhaps the strongest statement possible at this juncture is that very little evidence was discovered to indicate that international tourism is obstructing international understanding in Barbados. Fears that "ugly American" tourists may be spreading ill will and generating international misunderstanding everywhere are not consistent with the findings of this study . . .

The impact of tourism on traditional life-styles is especially important where those traditions form the basis for the development of tourism. Ethnic tourism is particularly a feature of the Third World but the arts in general have been, and still are, a very significant part of Europe's appeal to the traveller. Basically two differing schools of thought and bodies of evidence exist (V. Smith, 1977b). Tourism is held by some observers to have a corrupting influence, brought about by the cheapening of artistic values or the commercialization of local traditions and customs; for example, the performance of religious or historical ceremonies on demand, out of context and for monetary reward. Other writers suggest tourists can have a strengthening and stimulating effect, either by reawakening interest in a society's own culture or simply by supporting the ballet, theatre or museums by their presence and entrance fees. Overseas tourists are estimated to have bought a third of all theatre tickets in London in 1977 (ETB, 1981). More generally, seasonal tourist demand and other business generated from tourism-related workers may support shops and services which might otherwise disappear from some rural and other communities (Duffield and Long, 1981). Tourism may also impose other more banal but no less significant pressures on host populations, as when visitors are served ahead of local residents in shops and the latter are jostled in the streets during the height of the season, or restricted from the use of certain beaches (Rothman, 1978; Garland and West, 1985).

The demonstration effect noted earlier also encompasses daily patterns of consumption. Emulating the visitors, the residents may adopt new clothing styles, begin eating and drinking the imported food and beverages favoured by the tourist, or aspire to obtain the transistor radios, cameras and other material goods so casually displayed by him. Inability by the local to emulate the visitor, if so desired, may lead to a greater awareness of poverty, envy, frustration and consequent ill-will towards the visitor or to his compatriots who have been more successful. At the same time, fulfilment of the desire for more imported goods may have significant adverse economic effects. This is not a one-way process, however, for on their return home, tourists may, for example, retain a taste for food and drink experienced abroad and seek out ethnic restaurants or food stores. McElroy and de Albuquerque (1986) argue, however, with respect to the Caribbean that the nature and extent of the demonstration effect, together with the processes by which it operates, have yet to be soundly documented by researchers.

Other impacts may also be felt in the generating regions (Table 6.10). Although vacations are spent at the destination, from the tourist's point of view many of the benefits may be felt most at home where a phase of eager anticipation precedes, and a period of fond recollection follows, the holiday itself. If tourism is seen as a means of meeting personal needs and resolving interpersonal problems (Ch. 4), then the fulfilment of these needs and any resolution of these problems through a successful holiday should be beneficial not only to the individuals concerned but also to the communities from which they are drawn.

Attitudinal and longitudinal studies

Within destinations, attitudinal surveys have been the common method of addressing a range of social/cultural impacts, with some surveys also encompassing economic and environmental issues. In this approach, respondents are asked to indicate their attitudes and feelings towards the impact of tourism, sometimes through the use of unprompted open-ended questions, more frequently with a set of closed questions or statements. Such surveys can yield useful information on types of impact but it must be remembered that what is being recorded is the respondent's perception of those impacts and these may or may not accord with reality. Pearce and Cant (1981) found Queenstown residents' views of increased employment opportunities generally reflected the trends revealed by labour force surveys. In Gwynedd, Wales, tourism's economic contribution was underestimated when compared to revenue generated by other sectors (Sheldon and Var, 1984). Such cross-checks, however, are often not undertaken and for many social/cultural issues there may be few objective controls.

Attitudinal surveys can be especially useful in distinguishing the impacts felt by different groups within the community, provided the sample is large enough to be disaggregated. Comparison of results is limited by different approaches and means of analysis but some common patterns do emerge. There are also signs of considerable variation from one case to another and much evidence of the complexity of the impacts which tourism may generate.

Studies in coastal resort communities in North America (Pizam, 1978; Rothman, 1978; Thomason, Crompton and Kamp, 1979; Keogh, 1982), in three Scandinavian resort towns (Haukeland, 1984), in Sri Lanka (Ahmed, 1984) and in Rotorua, New Zealand (Garland and West, 1985) indicate that attitudes vary with the degree of involvement in tourism – more positive and fewer negative impacts are expressed by those economically dependent on the tourist industry. However, Liu and Var's (1986) research on attitudes in Hawaii revealed few differences between those working in tourism and in other sectors; this they attributed to the maturity of Hawaii as a destination.

Analysis of the attitudes of three community groups – tourism entrepreneurs, public providers/administrators and residents – has been undertaken in Corpus Christi, Texas (Thomason, Crompton and Kamp, 1979) and in three different types of tourist centres in England (Murphy, 1983). The reaction of the three groups in each case was reasonably favourable, with the business sector again generally being the most positive and the residents having more reservations. In the American case the administrators felt winter visitors placed strain on community services while in the British centres they were encouraged by the prospect of additional facilities. Murphy concluded, however, that the differences between the groups in the communities he studied were sufficiently close that they could be overcome.

Socio-economic status was reported to have little effect on attitudes to tourism amongst residents of Santa Marta, Colombia (Belisle and Hoy, 1980), nor on acceptance of foreign visitors in West Virginia (J. Pearce, 1980). However, in a survey of three New Zealand resort towns, Garland and West (1985) found (p. 38): 'Younger residents (18–29 years), Maori residents, women and lower-class residents were less enthusiastic about the role of tourism in the community'. There was also less support from those without overseas travel experience. Some of these findings contrast with those reported by Keogh (1982) in Shediac, New Brunswick where the young generally had a favourable attitude as the result of the prospect of summer jobs and greater social contact with tourists.

Ahmed (1984) assessed the attitudes both of local residents (English-speaking, middle or upper class) and foreign tourists in Sri Lanka and found the two groups shared reasonably similar views on many of the economic and social consequences of tourism in that country, with Sri Lankans showing more concern for prostitution and drug-use than the visitors.

Other research has considered resident attitudes to differing groups of tourists. Sheldon and Var's (1984) analysis indicated residents of North Wales held stereotyped views of visitors, saw those from more distant places as more considerate and felt the more frequent visitors, with the exception of the English, had the least negative impact on their society (no account was taken of Welsh tourists). A similar pattern is reported from Turkey where tourists from more distant countries were ranked more favourably by the residents of Marmaris than visitors from the Middle East, Turkish workers from abroad and domestic tourists (Var, Kendall and Tarakcioglu, 1985). Reaction to overseas visitors in the three New Zealand communities studied by Garland and West (1985) was generally favourable but there was greater support for an increase in domestic visitors.

Useful insights into the relationship between perceived impacts and processes of development are provided by those studies which analyse attitudes in terms of exposure to different degrees of tourist development. Some researchers have focused on spatial variations within resorts while others have undertaken comparative studies of different resort areas. On Cape Cod, Pizam (1978) found more negative attitudes were expressed by mid-Cape residents, that is by those having the greatest contact with tourists. The opposite occurred in Santa Marta and Shediac where the impact of tourism was perceived less positively as distance from the tourist zone increased (Belisle and Hoy, 1980; Keogh, 1982). Keogh also noted a tendency for more negative attitudes amongst the more recently arrived residents of Shediac.

Haukeland's (1984) analysis of three small Scandinavian communities indicates a more negative attitude towards continued growth in tourism in the most developed areas. Likewise, in the Highlands and Islands of Scotland, Duffield and Long (1981, p. 423) note that:

> it was in the areas where tourist development was at its most embryonic that the perceptions of the economic benefits were high, and acknowledgement of the drawbacks of tourism development was least. In contrast, while the more developed tourist areas were no less aware of the economic benefits, they had a greater appreciation of the disadvantages that tourism can bring.

Conversely, in the New Zealand study (Garland and West, 1985) attitudes to tourism were generally more positive in the largest tourism centre, Rotorua, than in the smaller communities of the Bay of Islands, Tauranga–Mount Maunganui. Similar findings emerge from North Wales where Sheldon and Var report from a small sample that attitudes to tourism were more favourable in areas of high tourist density.

The results of these different studies suggest that as development increases there is no consistent firming of support for nor strengthening of opposition to tourism. The relationship between the level of development and attitudes towards tourism is clearly not unidirectional as implied in Doxey's (1975) iridex. Doxey's iridex is proposed as a 'gauge of irritation intensities' marked by progressive change through four stages: euphoria, apathy, annoyance and antagonism, as tourism develops. While there is support for this sequence in some of the case studies just outlined and in parts of the Caribbean (Ch. 3), other case studies and examples indicate that an antagonistic community is not the inevitable outcome of the later stages of tourist development. As Garland and West (1985, p. 35) observe in Rotorua: 'Host irritation with the presence of tourists in Rotorua is just that – irritation, and then only among small proportions of residents.' In other long-established tourist regions such as the South of France and parts of Switzerland, there appears to be a general acceptance of tourism and little of the antagonism suggested by Doxey's iridex.

Attitudinal surveys in resorts of different sizes and stages of development can provide some basis for exploring these issues, particularly if weight is given to the context in which the surveys are undertaken. Rothman (1978, p. 12) for example, concluded with reference to the two Delaware coastal resorts he studied that:

> . . . while vacationers have a significant impact upon the community, the impact does not appear to be disruptive. Rather, it seems to be a situation in which the residents are able to make a major adjustment with relative ease. This is probably because these communities have had long experience with vacationers and have been able to develop mechanisms of accommodation [e.g. by changing patterns of behaviour]. Consequently, they are able to cope with very large and heterogeneous populations without generating destructive conflict.

Husbands (1986) also underlines the significance of the time scale of development and notes (p. 180):

> . . . with the tourism industry well established and Barbados well known as a winter resort, it is certainly possible that residents and tourists have accommodated themselves to any stress which may be associated with relationships between tourists and residents.

He also emphasizes (p. 186) the role of spatial behavioural patterns, with tourists and residents largely frequenting different parts of Barbados:

> . . . such a division of places . . . , with the concomitant low levels of impact is not a situation which is consistent with a high probability of stress and conflict.

England, Gibbons and Johnson (1980) examined changes in subjective well-being in six ski-resort and six non-resort communities

Plate 11 Counteracting negative attitudes towards tourism: a roadside sign in Puerto Plata, Dominican Republic, urges residents to protect tourism as it increases sales and generates more jobs

in the intermountain west of the USA and found (p. 343):

> People appear to respond to increased formalization and impersonality by seeking voluntary associations and closer informal ties. They respond on the value level by reaffirming the importance of religion, loyalty, kindness and self-control, rather than accepting the less personal and less traditional values. This suggests that people may possess homeostatic mechanisms that set into action processes to counteract the alienating and depersonalizing effects of modernization.

These authors conclude their study by noting (p. 346):

> . . . ski resorts have impacts on political and work alienation, slightly increasing both. They also seem to account for increases in present well-being, future well-being, and change in well-being from the past to the present. The results fail to provide support for those who depict ski resorts as powerfully positive contributions to the well-being of residents in nearby communities, although there are slight increases in most measures of well-being. On the other hand, alienation does increase.

England, Gibbons and Johnson assessed these changes by asking respondents to rate their lives 'at present, five years ago and five years hence'. There are, however, limits to which attitudinal surveys can be carried out retrospectively and no attempt appears to have been made yet to repeat any of the surveys undertaken at an earlier date. Consequently longitudinal studies in the same community or resort may provide even better insights into how impacts and attitudes change as tourism develops than a series of comparative snapshots at a given point in time provided by the studies outlined above. This is particularly so with those social/cultural impacts which can be more readily measured than changing attitudes as White (1974) has demonstrated with his study of language change in Switzerland and as Getz (1986a) has shown with his analysis of demographic trends in Scotland.

A particularly useful example of a longitudinal study is Préau's (1983) analysis of change in Saint-Bon-Tarentaise (Savoy). Préau provides much detailed contextual material, outlines clearly the type of development which occurred (Courchevel was a planned ski-field, a precursor to the integrated resorts of the 1960s) and traces household changes from baseline dossiers established by the development agency in 1948 and his own field work in 1972–73. As noted earlier, this approach enabled Préau to examine in detail the degree and type of participation of a sample of households in the development process and to analyse the occupational and social changes which occurred from one generation to the next. The existence of the baseline dossiers in Saint-Bon-Tarentaise is somewhat exceptional, but consideration might be given to establishing such data when new projects are proposed elsewhere, so that subsequent social and cultural impacts might be monitored as development proceeds.

ASSESSING THE ENVIRONMENTAL IMPACT OF TOURIST DEVELOPMENT

Assessment of the environmental impact of tourism is particularly important for various environmental factors constitute the basis of much tourist development (see Ch. 5). Moreover, tourists tend to be attracted to some of the more complex and fragile environments, for example, small islands, coastal zones, alpine areas and centres of historical or cultural interest. Jackson (1984, p. 7) observes with reference to the Caribbean:

> Some tourism facilities, such as marinas, are water-dependent, so that they straddle dynamic and highly vulnerable littoral zones. Others, such as beach hotels, though not necessarily water-dependent, view accessibility to beaches as a significant plus. Marinas are inclined to favour lagoons that are often characterized by productive associations, in regard to fisheries, mangroves, seagrass and reefs.
>
> The result of these factors is that the majority of Caribbean tourism facilities are sited within less than 800 metres of the high-water mark, in a zone that can be both unstable and vulnerable to geological, oceanographic and meteorological phenomena.

In assessing the environmental impact of tourism, as with its economic evaluation, account must be taken of the composite nature of tourism. Particular impacts need to be related to specific aspects of this multi-faceted activity but at the same time some broader synthesis must be retained (Pearce, 1985b). Most studies so far have focused only on one type of impact or aspect of tourism but several useful disaggregate approaches have been put forward. In their review of ecological problems in island ecosystems McEachern and Towle (1974) differentiate between aesthetic, ecological and cultural values and provide a framework which identifies potential impacts, measured in terms of their magnitude and importance, associated with a range of major land-use activities. In the more specific case of water-based recreation, Seabrooke (1981) outlines a framework which identifies different recreational activities, the areas affected by each and the nature of the effects arising. Liddle and Scorgie (1980) differentiate between physical and chemical impacts resulting from different shore-based and water-based activities. A particularly comprehensive framework (Table 6.11) was developed to ensure comparability between the case studies prepared in 1977 and 1978 as part of the OECD's tourism and environment programme (OECD, 1981a, 1981b). Table 6.11 identifies a number of tourism-generated stressor activities, the associated stresses, subsequent environmental responses and man's reactions to these, both individually and collectively. Stress has been defined as: 'the strain imposed on people and their enjoyment of amenities or

Table 6.11 A framework for the study of tourism and environmental stress.

Stressor activities	Stress

1. *Permanent environmental restructuring*
 (a) Major construction activity
     urban expansion
     transport network
     tourist facilities
     marinas, ski-lifts,
     sea walls
 (b) Change in land use
     expansion of
     recreational lands
	Restructuring of local environments
	     expansion of built
	     environments
	     land taken out of
	     primary production

2. *Generation of waste residuals*
     urbanization
     transportation
	Pollution loadings
	     emissions
	     effluent discharges
	     solid waste disposal
	     noise (traffic, aircraft)

3. *Tourist activities*
     skiing
     walking
     hunting
     trial bike riding
     collecting
	Trampling of vegetation and soils
	 Destruction of species

4. *Effect on population dynamics* Population growth	Population density (seasonal)

(*Source:* after OECD).

Table 6.11 continued.

Primary response environmental	Secondary response (reaction) human

Change in habitat
Change in population of
  biological species
Change in health and
  welfare of man
Change in visual
  quality
	Individual – impact on aesthetic values
	Collective measures
	  expenditure on environmental improvements
	  expenditure on management of conservation
	  designation of wildlife conservation
	    and national parks
	  controls on access to recreational lands

Change in quality of
  environmental media
  air
  water
  soil
Health of biological
  organisms
Health of humans
	Individual defensive measures
	Locals
	  air conditioning
	  recycling of waste materials
	  protests and attitude change
	Tourists
	  change of attitude towards the environment
	  decline in tourist revenues
	Collective defensive measures
	  expenditure of pollution abatement
	    by tourist-related industries
	  clean-up of rivers, beaches

Change in habitat
Change in population
  of biological species
	Collective defensive measures
	  expenditure on management of conservation
	  designation of wildlife conservation
	    and national parks
	  controls on access to recreational lands

Congestion
Demand for natural
  resources
  land and water
  energy
	Individual – Attitudes to overcrowding
	  and the environment
	Collective – Growth in support services,
	  e.g. water supply, electricity

on resources, the impact of which can be objectively measured or may be subjectively experienced in the light of defined values'. The emphasis here is on measurable stress and on linking specific stresses and responses to particular activities. Jackson (1984) presents a similar framework in which the major columns are the area of impact, the source (use, activities) and the nature of the impact (direct environmental and related socio-economic).

The first major source of environmental stress identified in Table 6.11 is permanent restructuring of the environment brought about by a variety of major construction activities such as new urban developments, construction of highways and airports and the building of recreational facilities, for example marinas or ski-lifts. The extent of these changes can be quite substantial, as in the ribbon development of France's Riviera, often characterized as a *mur de béton*. Likewise the Belgian coast, with over twenty resorts located along its 67 km, has been developed to some degree for tourism for virtually its entire length (Pearce, 1987a). Figure 3.1, which is based on aerial photograph interpretations, depicts the changes that have occurred in the built environment of Lloret de Mar on Spain's Costa Brava (see Ch. 3) from 1956 to 1981 (Priestley, 1986). During this period the urban centre has trebled in size and *urbanizaciones* have come to cover a quarter of the commune's total surface area with a consequent transformation of former horticultural and forested land (Plate 12). Further south, Joliffe and Patman (1986, pp. 120–2) catalogue the impacts which have occurred in Roses Bay as a result of environmental restructuring and other tourism induced changes:

> Potential impacts are numerous and varied, and include: land and sea pollution, loss of wetland, loss of agricultural land and changes in the structure of agricultural demand, changes in shoreline configuration, loss of sediment in some areas, siltation and dredging requirements in others, grossly sub-optimal public benefits in the coastal environment, sand encroachment into the amenity areas, coastal flooding (a 70 year event?), increasing coast protection needs, macro and micro access problems, eutrophication, water abstraction that leads to sediment impoundment and impact on shoreline sediment budget.

Aspects of the environmental impacts of large recreational subdivisions in the USA, the American counterpart to the Spanish *urbanización*, are outlined by Stroud (1983). He notes that severe erosion may result from a prolonged development period, with 'premature' subdivision leaving access roads unpaved and exposed to the elements for a number of years. However, even dispersed second home development has its impacts, as in Canada where Racey and Euler (1983) estimate habitat disturbance by lakeshore cottages to be constant at 0.1333 ha per cottage, regardless of the size of the lot.

Other research has examined aspects of environmental restructuring

Plate 12 The built environment: intensive development along the waterfront at Lloret de Mar, Spain, has been compounded by the construction of villas on hillslopes overlooking the town. The palm-lined promenade and plaza also testify to the 'tropicalization' of the resort. (Photo courtesy of Gerda Priestley)

brought about by the development of the attractions sector. Chmura and Ross (1978) provide a comprehensive review of the impacts, both positive and negative, of marinas and their boats, together with a discussion of management implications. They note (p. 4):

> The primary negative impacts are habitat loss, pollution by stormwater runoff, and aesthetic (visual) pollution. A marina's impact can also have positive features, since it provides for the concentration of shoreline development (as opposed to many scattered private docks) and may increase the diversity of shoreline habitat, e.g. providing substrate for fouling communities.

Consideration is also given to the effects of dredging and the construction of breakwaters, piers and wharves. Elsewhere, Baines (1987) observes that some coastal restructuring may not be as permanent as intended, with some 'quick-solution' groynes designed to trap sand and build up beaches on small sand cays actually inducing significant coastal erosion. Serious losses of beach area in Barbados are partially attributed to increased erosion resulting from damage to coral reefs caused by a variety of recreational activities and blasting of access channels (Archer, 1985). Damage to coral reefs, mangroves and other marine habitats can also result from increased sedimentation during the construction phase.

In alpine areas, erosion may result from the construction of ski-lifts and terrain modifications undertaken to improve the trails (Candela, 1982; Mosimann, 1985). Considerable variation may occur in the extent of the erosion due to differences in the modifications made and to local factors. In a review of Swiss ski-fields, Mosimann found the main factors determining the risk of erosion were slope form, soil moisture status, frequency of runoff and size of the catchment. Depending on the altitude of the resorts, significant deforestation may result from the clearing of ski-trails and construction of lifts. In resorts such as Vars, up to 30 ha have been cleared (Candela, 1982). Deforestation in this way may result in avalanches, gullying and hydrologic changes. In some cases, man's response to the stress created may subsequently alleviate some of these problems. Bayfield (1974), for example, notes the decline in accelerated erosion resulting from the construction of new chair-lifts on Cairngorm after the damaged ground was reseeded and drains were provided.

In other instances, white elephants testify to the permanence of structures built for a demand which never existed or failed to materialize. This problem is well illustrated by the ill-conceived tourist developments at Samaná in the Dominican Republic. Despite the personal support and interest of a former president (Symanski and Burley, 1973) and an attractive site, Samaná has been handicapped by difficulties of access and the project has been unsuccessful. Shops

Plate 13 The built environment: new shops at Samaná, Dominican Republic, remain unopened long after the new tourist complex was constructed on a site cleared of its original residents.

constructed on a site cleared of its original dwellings have remained unopened (Plate 13) and a massive footbridge linking small islands to the main hotel is little used.

A second major area of stress is that resulting from the generation of new or increased waste residuals. One of the most interesting studies on air pollution in resort communities is that by Kirkpatrick and Reeser (1976) on Aspen and Vail (Colorado). Their study emphasizes the importance of context in impact studies. It was demonstrated that mountain altitude and terrain features in Vail and Aspen seriously inhibited air pollution dispersion as compared to Denver and the Colorado Plains. Moreover, automobile emissions were found to be higher in mountain communities (as a result of the effects of altitude and slower travelling speeds), as were particulate emissions due to the large-scale use of open fireplaces for heating and social effect. As a result of these factors it was shown that for similar populations, in terms of carbon monoxide, the air quality at Vail is about ten times as fragile as that in Denver.

However, in terms of waste residuals, the most widespread problem in resort communities is water pollution through the discharge of inadequately treated effluent. Seas, lakes, rivers and other water bodies which are amongst the most attractive resources for tourist development are also frequently used for the cheap and convenient disposal of sewage. This practice may in time give rise to the eutrophication of these water bodies through an increase in discharged phosphates or contamination such that human health may be seriously impaired and/or natural flora and fauna destroyed. The 'collapse' of the Millstatter Lake in Austria during the early 1970s following a tenfold increase in the tourist traffic in the preceding two decades and restrictions on bathing in certain Mediterranean beaches, are cases in point. These examples also emphasize the tourism/environment relationship. Whereas industrial discharges, for example from chemical plants, would have little effect on that industry, closure of beaches or serious changes in the aesthetic qualities of lakes may result in a significant downturn in the tourist traffic. Although the fragility of these environments is important, the problem here is basically one of management and stems essentially from an inadequate infrastructure (Ch. 5). This may result from the rapid expansion of tourism wherein construction of accommodation outstrips the provision of treatment facilities. More commonly it arises simply from a basic inability or unwillingness of local authorities to finance the necessary sewerage plant. Once this financial handicap is overcome the situation can usually be reversed and this source of stress considerably reduced. Economically, as well as environmentally, provision of adequate infrastructure from the outset will in the long run prove less expensive than the correction of environmental damage plus the eventual construction of the necessary plant.

The best documented aspect of environmental impact is that concerning recreational activities, although most of these studies refer more to picnic grounds, national parks and wilderness areas rather than to resorts as such (Wall and Wright, 1977; Edington and Edington, 1986). Many of these have been concerned with the trampling effects on soils and vegetation by various activities such as skiing (Baiderin, 1978; Candela, 1982), off-road vehicles (Crozier, Marx and Grant, 1978; Kay *et al.*, 1981), and walking (Liddle, 1975), in a range of environments including coastal ecosystems (Ghazanshahi, Huchel and Devinny, 1983), dunes (Bowles and Maun, 1982; Lundberg, 1984; Brown *et al.*, 1985), and forests and meadows (Dale and Weaver, 1974; Cole, 1987). A comprehensive review of the impacts of different activities – camping, skiing, horse-riding and hiking – in North American alpine areas is given by Price (1985). The effects of trampling include an increase in soil compaction and erosion and changes in plant cover and species diversity. Other research has focused on the impacts of sightseeing (Pedevillano and Wright, 1987), collecting (Ghazanshahi, Huchel and Devinny, 1983), and boating (Chmura and Ross, 1978; Liddle and Scorgie, 1980; Johnstone, Coffey and Howard-Williams, 1985). In addition to a wide-ranging review of the environmental impacts of such activities, Edington and Edington (1986) also consider tourism-related disease hazards, insect nuisances and hazards associated with larger animals.

A fourth associated area of impact is the effect of tourist development on population dynamics, especially seasonal increases in population and population densities. One of the more obvious effects of such seasonal increases is the resultant physical congestion experienced in many areas, be they beaches, ski-slopes or historic centres. Overseas tourist arrivals in New Zealand, for example, are highest from October to March, which coincides with New Zealand's peak domestic holiday season. However, in some urban areas the effect of the influx of tourists may be lessened by the outflow of local holidaymakers, for example as several million Parisians head for the coast or countryside in July and August. Seasonal influxes will also increase the demand for natural resources such as water and energy and contribute to some of the effects already noted; for example, the generation of waste residuals.

In some instances, the adverse impacts of tourist development may be negligible. Reviewing the environmental impact of tourism in the Khumbu or Everest region of the Himalayas, Pawson *et al.* (1984, pp. 244–6) conclude:

> The available evidence suggests that the growth of tourism in Khumbu during the past 12 years has occurred without significant deterioration of the region's delicate environment, except in a few locations where waste disposal or excessive grazing present major problems. The very extensive

deforestation in many parts of what is now the Sagarmatha National Park appears to have originated long before the onset of the tourist influx in the early 1970s. . . . Except in Namche Bazar, tourist-related construction has not altered the appearance of human settlements, and even in Namche most new buildings are constructed to blend with existing architectural styles.

The level of impact appears to be related to the volume of tourist traffic (about 5000 trekkers and other visitors per annum), the scale of the developments and controls such as the visitor fuel-use regulations.

Moreover, tourist development does not necessarily give rise solely to adverse environmental impacts (Cohen, 1978; WTO, 1983d). Much development, for the tourist at least, may enhance his appreciation of the environment. The construction of roads or cableways will give him greater access to viewpoints or open up new ski-fields; the provision of accommodation facilities will enable him to stay in the area. More generally, tourism may be the means of preserving areas of scenic beauty or centres of historical interest by providing an economic or social rationale to reinforce purely environmental or historical considerations which have often proved insufficient by themselves in the past. Studies of the economic impact of visitors to New Zealand national parks, for example, appear to have been undertaken largely to indicate that areas set aside from extractive resource use can still contribute significantly to the local, regional and national economy (Pearce, 1982; Entwistle, 1987). In Ch. 3, it was shown that tourism often provided the economic rationale for a variety of urban restoration and renovation projects. Chmura and Ross (1978, p. 5) point out with regard to the aesthetic effects of marinas, 'it may be assumed that a marina situated on a pristine shoreline will have a negative effect, while one placed on a developed or urban waterfront may actually improve the appearance and environmental quality of the waterfront'. They also note that the impacts of dredging are not always adverse and that it may bring positive ecological effects, namely: 'It may help improve circulation in choked inlets, increase the availability of food to fish and shellfish and help to flush and dilute polluted waters.'

Some of these beneficial environmental effects may be indirect and unintended; in other instances changes have resulted from an explicit attempt to modify or improve the environment. Such is the case with the introduction of tropical ornamental vegetation to the French Riviera during the nineteenth century (Gade, 1986). The initial success and subsequent spread of this trend to other Mediterranean regions and abroad has been such that the palm-lined avenue or square is now a characteristic feature, albeit an exotic one, of many coastal resorts throughout the world (Plate 12). According to Gade (p. 6), 'The pseudo-tropical landscape has fostered tourism to the region [the

Riviera] by its power to suggest *dépaysement* . . . the sensation of being in a different kind of place from that usually experienced.' Thus physically changing places [for the better?] reinforces the change of place factor underlying tourist demand discussed in Ch. 4.

The comprehensiveness of the approach outlined in Table 6.11 is appealing, but the demands and complexities of undertaking any particular step in it should not be underestimated (Pearce, 1985b). In many cases an inadequate understanding of tourist behaviour and recreational activity is compounded by incomplete knowledge of environmental processes. Many of these problems were highlighted by Turner (1977) when he attempted to incorporate a water quality measure in a recreation demand model. Firstly, there was the question of what to measure. Here Turner cites Freeman (1975) who noted: 'water quality cannot be represented by a single number on some scale but rather is an n-dimensional vector of the relevant parameters. Further, water quality varies across space and time, making the task of measurement and description even more complicated.' Secondly, different factors are likely to affect different recreational activities in different ways, such that a change in one parameter, for example water temperature, might benefit an activity such as swimming but adversely affect another, for example trout fishing. It is also not yet very clear what water quality factors are perceived by recreationists to be important, assuming it is their perception of water quality rather than the actual conditions which determine their behaviour. Where there are a variety of recreational activities or where tourism is but one part of the economy, identification of the source of any deterioration of water quality is likely to be more complicated. Edington and Edington (1977) caution against the tendency to attribute any loss of environmental quality to the most visible cause and cite the case of Llangorse Lake in South Wales where effects attributed to speed boats appeared on closer inspection to be more likely a consequence of inadequate sewage treatment at a nearby camping ground. It should also be borne in mind that it is not only recreational and other activities which may vary from one water body to another for natural nutrient variation may occur to complicate comparative studies (Silverman and Erman, 1979).

Such problems are not restricted to water-based recreation. In their detailed study of visitor impacts in Yosemite National Park, for example, Foin *et al.* (1981) note 'There is considerable variation between sites . . . much of which could probably be traced to site factors. In more general terms, the compounding of visitors, impacted populations, and site factors is a serious problem that to date has not yet been faced.' Other complicating factors noted by these and other researchers include the existence of indirect and secondary effects, seasonal variations, the possibility of time lags and differing recovery rates (Wall and Wright, 1977; Bayfield, 1979; Joliffe and Patman, 1986). Habitat changes, for example, can subsequently affect wildlife, and together with other recreational effects such as disturbance and destruction may set in train a further series of impacts (Fig. 6.5).

Three major sets of techniques for analysing these various impacts in detail may be identified: after-the-fact analyses or analytical studies, simulation or experimental methods and the monitoring of change through time (Wall and Wright, 1977; Price, 1985). Each has associated problems and advantages and may be more appropriate for some types of studies than others. After-the-fact or analytical studies, for example, require a sound knowledge of conditions prior to development or access to adjacent undisturbed or unmodified areas. Wall and Wright suggest that such an approach is more suitable for soil and vegetation studies than for those of water quality and wildlife where it may be more difficult to establish controls. Aerial photograph interpretation can be a particularly useful means of examining changes in the built environment as Priestley (1986) has shown (Fig. 3.1). Experimental studies have been undertaken most frequently with regard to trampling (Price, 1985). Monitoring can be very useful, particularly in allowing the simultaneous study of both cause (tourist development) and effect. Good examples of this approach are provided by the *Coral Reef Monitoring Handbook* published by the South Pacific Commission (Dahl, 1981) and Greenland's (1979) review of modelling air pollution potential in mountain resorts. Certain types of impact are more readily measured

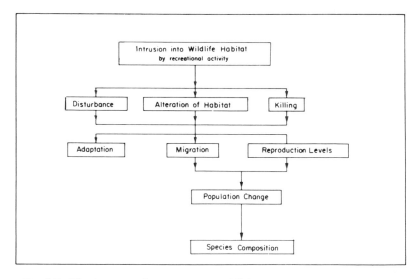

Fig. 6.5 The impacts of recreation on wildlife.
(*Source:* Wall and Wright, 1977).

than others. Atmospheric particulate levels or coliform counts can be established rather more easily than levels of visual pollution.

Much scope exists for developing each of these approaches further. Attention, however, must be directed not only at furthering our understanding of the physical processes at work and the ecological consequences of these but also at clarifying the linkages between tourist development and these changes and measuring their extent and spatial occurrence (Pearce, 1985b). In this respect, the Lloret de Mar and Roses Bay studies discussed earlier are especially helpful as they not only discuss the processes but also delimit the location and extent of many of the changes (Fig. 3.1).

The OECD programme mentioned above yielded valuable insights into processes of tourist development and associated environmental impacts (OECD, 1981a, 1981b). Analysis of the seventeen case studies and eleven national reports indicated environmental deterioration occurred most commonly under the following circumstances: a rapid and largely uncontrolled growth of the tourist industry, especially where this is characterized by marked seasonal peaks, in areas with little or no planning, few controls, and financial or technical inability to provide adequate infrastructure. Decreases in the tourist traffic attributable to a loss of environmental quality were difficult to measure but were shown to be more common where a single environmental attraction was dominant. For example, an inland lake area suffering from eutrophication experienced twice the rate of decline in tourist numbers as the surrounding State. Elsewhere, McEachern and Towle (1974) suggest a decrease in the tourist traffic is not an inevitable consequence of environmental deterioration:

> As the character of the tourist island changes from 'unspoiled' to 'spoiled' (e.g. more urbanised), the influx of tourists does *not* stop. Due to various stimuli, such as mass advertising and other publicity, it continues and may even accelerate according to a positive feedback loop that reinforces the degree of urbanisation and the influx of new tourists with different tastes.

The economic impact of decreases or changes in the tourist traffic attributable to environmental degradation have yet to be firmly established.

Although the OECD programme provided little evidence of policies with no negative environmental impacts, various measures (zoning, building and land use controls, provision of infrastructure) were shown to reduce adverse impacts significantly (see also Inskeep, 1987). The costs of such measures were not fully analysed in the case studies but the inability of many communities to meet these charges was clearly the source of many of the problems experienced. The case of Queenstown, New Zealand, illustrates some of the issues that may arise (Pearce, 1978c). During the 1960s the growth of tourism placed stress on the town's existing sewerage system, with effluent discharge polluting a branch of Lake Wakatipu. Financial assistance from central government for a new sewerage scheme was eventually made available when the construction of new hotels was delayed by the refusal of the local authority to issue building permits. The new scheme removed the threat of lake pollution from this source but resulted in a higher degree of local authority indebtedness and a change in the rating system which has passed more of the burden on to the tourist industry. Subdivision and sale of lands on nearby Queenstown Hill has contributed to the repayment of these loans but the subdivision itself involves sensitive environmental considerations.

Many of the other papers cited above conclude with management implications and recommendations for alleviating environmental stress. The WTO (1983d) attaches great importance to spreading tourist demand throughout the year so as to reduce seasonal pressures upon the environment, particularly in resort areas. At the same time, their report notes that a growth in short holidays, particularly weekend tourism, may generate increased pressures on peri-urban areas through the spread of motorways, increased vehicle emissions and loss of some peri-urban open spaces. They also advocate the construction of small hotels and rented or co-operatively owned accommodation rather than expansion of second homes so as to avoid the breaking up of the landscape and privatization of rural space, excessive seasonal concentration of demand and an increase in infrastructure. Whether such a policy is feasible at a time when the trend in many places is towards self-catering is debatable. Other aspects of the incorporation of environmental issues in planning will be examined further in the following chapter.

CONCLUSIONS

The literature and issues reviewed in the preceding sections clearly show tourist development has many and varied impacts. Whether the focus is on economic, social/cultural, environmental or all three issues, examples can be found which tend to stress more positive or more negative aspects be they concerned with inter-sectoral competition, female employment, attitudes to tourism or the impact on natural landscapes. That such variety should be found is not surprising given the spectrum of processes reviewed, the different sorts of contexts discussed, and the range of techniques employed and the perspectives adopted by researchers from the many disciplines which have addressed, directly and indirectly, these issues. For these reasons an attempt has been made to structure the examination of these impacts in terms of the major frameworks presented (Tables 6.1, 6.10, 6.11 and Figs. 6.1 and 6.2).

Such frameworks provide useful integrative devices by stressing the interrelationships between various sets of impacts, identifying present gaps in our knowledge and suggesting lines for further research. Two major concerns emerge here; strengthening the linkages between studies of different sorts of impacts and refining particular techniques further. The first provides challenges to the generalist or to interdisciplinary teams; particularly the bringing together of those impacts measured in quantitative terms with those expressed in a more qualitative manner. The second calls for further specialist involvement in improving techniques and in their application, be it the refinement of calculations of the balance of payments effects of tourism, longitudinal studies of social impact or the monitoring of environmental change. Attention in all of these areas must also be directed at evaluating the performance of tourism against that of other sectors if tourism's contribution to development is to be fully assessed. These issues are examined further in the reappraisal of tourism and development in Ch. 8.

The following chapter (Ch. 7), considers some of the ways in which planning can reduce the adverse impacts of tourist development and increase the benefits which might arise from it.

Planning for tourism

In Chs. 1 and 2, tourism was shown to be a diffuse and complex activity, consisting as it does of a wide range of elements which may be developed in a variety of contexts by a broad spectrum of developers, each having different aspirations and capabilities. Chapter 4 highlighted the diversity of tourist demand while Ch. 5 outlined the multiple resources which might be developed for tourism. Chapter 6 showed tourist development involves not only tourists and developers but also touches other sectors of society, is directly related to the economy in general and may impact on the environment at large. In particular, problems arise and costs are increased when the different tourism sectors do not develop harmoniously or when the motives and capabilities of the different development agents conflict. Carried to an extreme, uncontrolled growth of tourism can destroy the very resource base on which it was built. Of the different development processes examined (Ch. 3), those involving some degree of planning were most often, though not always, shown to reduce the externalities of tourist development and to enhance its positive impacts (Ch. 6).

Recognition of the importance and value of planning for tourism is reflected in the number of tourism plans which have been prepared in the last two to three decades. The World Tourism Organization established an inventory of over 1600 assorted tourism plans in 1980 (WTO, 1980c). Different scales of tourism planning were identified by the WTO: intraregional, national, regional, local and sectoral (concerning one or more sectors of tourism, e.g. social tourism, coastal or mountain areas). Moreover, at each scale different types of plans have been formulated. At the national level these may include:

1. General national plan – a national development plan including tourism.
2. National infrastructure plan – a plan establishing guidelines for the development of infrastructure at the national level, including tourism.
3. National tourism development plan – a specific plan for the development of tourism at the national level.
4. Tourism infrastructure plan – a plan establishing guidelines for the development of tourism infrastructure at the national level.
5. National promotion and marketing plan – a plan or programme of promotion and marketing of tourist products at the national level.

A similar range of plans may be found at the other scales so that in any one country tourism planning may take many shapes and forms as Heeley (1981, p. 61) graphically points out in the case of Great Britain:

> The scope of existing arrangements for planning for tourism is an amalgam of economic, social and environmental considerations. Three distinct geographical planning levels (region, county and district) are discernible; a large number of public agencies is active at each level. Overall planning objectives in tourism vary from encouragement to restriction, while planning tools range from grants and other incentives to development control schemes and visitor management projects . . . Moreover, tourism is not a self-contained policy area. It overlaps with policy fields such as transport, conservation, rural development, and so forth, so that only a small proportion of the sum total of plans affecting tourism are exclusively devoted to it.

APPROACHES TO TOURISM PLANNING

Variations in approaches to tourism planning add to the complexities of scale and different types of plans. Moreover, there has been a significant evolution in approaches to tourism planning as recent reviews have highlighted (Braddon, 1982; Acerenza, 1985; Baud-Bovy, 1985; Murphy, 1985; Getz, 1986a, 1987). In general, these writers identify or advocate a move away from a narrow concern with physical or promotional planning facilitating the growth of tourism to a broader, more balanced approach recognizing the needs and views of not only tourists and developers but also the wider community. Concern is now expressed that tourism should be integrated with other forms of social and economic development. In these respects, the evolution of tourism planning parallels and is related to the changing attitudes to development discussed in Ch. 1.

Getz (1987, p. 3) defines tourism planning as:

A process, based on research and evaluation, which seeks to optimize the potential contribution of tourism to human welfare and environmental quality.

According to Murphy (1985, p. 156):

Planning is concerned with anticipating and regulating change in a system, to promote orderly development so as to increase the social, economic and environmental benefits of the development process. To do this, planning becomes 'an ordered sequence of operations, designed to lead to the achievement of either a single goal or to a balance between several goals' (Hall, 1970 p. 4).

However for Braddon (1982, p. 2):

1. ... There is no single definition of tourism planning so an analysis of its constituent elements helps an understanding of the term ...
2. Tourism planning is very closely linked with development planning in most parts of the world. The planning process needs to take account of very many factors ranging from topography to economy and from tourists' needs to residents' needs. It is subject to a great many external influences which both modify the process and the outcome – the implementation of the plan.
3. Tourism is a social, economic and environmental activity. Its planning has to operate at various levels; nationally, regionally, locally.
4. Tourism planning must take account of conservation of the physical environment.... The spatial planning of tourism can be very effective in this regard.
5. Ideally tourism planning should be fully integrated with all socio-economic activities and at all levels of involvement. This would ensure the optimal use of tourist resources with least social, economic and environmental costs.
6. Tourism planning is not just the formulation of plans for the future. It is also about the implementation of plans. It is therefore important that the right economic conditions exist for development to take place in accordance with the plan. It is also important that tourism planning is market-oriented, providing the right product for the consumer – the tourist.
7. At the outset of planning the aims must be stated ...

Many of these points are also advocated by Acerenza (1985) who applies a strategic planning approach to tourism. Figure 7.1 depicts the basic stages involved in this longer term, administrative approach to tourism planning. In particular, Fig. 7.1 shows planning begins at a much earlier stage than the determination of the development strategy which was the prime focus of many earlier tourism plans. For Acerenza, the process begins with an analysis of what has already been achieved, that is with a critical assessment of the various impacts, both positive

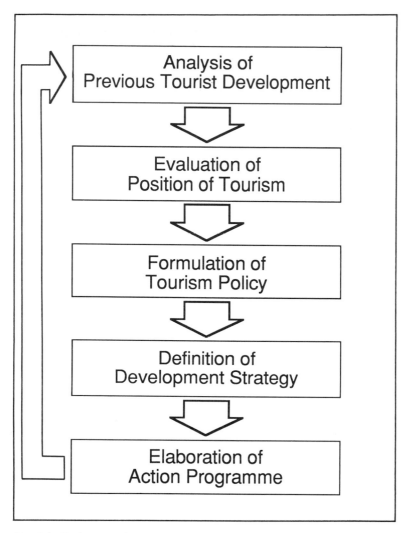

Fig. 7.1 Basic stages in tourism planning.
(*Source:* Acerenza, 1985).

and negative, of previous tourist development and an analysis of the political significance of tourism.

This information is fed into the second stage, one which Acerenza sees as being fundamental but one which was often omitted or given little weight in earlier plans, that is an evaluation of tourism's position from different perspectives. Firstly, tourism must be seen in terms of national social and economic priorities so that it might contribute more effectively to meeting broader development goals and so that the level of support from central government for tourism, and the implementation of any tourism plan might be gauged. Such an evaluation requires tourism to be set against other sectors but as the previous chapter has indicated, few impact studies have so far attempted to do this. However, Ch. 6 did show that research into community attitudes towards tourism is reasonably well developed, although incorporation of such views into the planning process is far less common. This is the 'community approach' advocated by Murphy (1985), an approach which would appear much more feasible at the local rather than the national level. Clearly the aims and expectations of the tourist industry itself must also be taken into account as much development will result directly from private sector initiatives and activities (Chs. 2 and 3). Particularly important here are the organizations which exist to articulate the many and varied interests of the different sectors and to channel these effectively to the appropriate planning agencies. Evaluation of tourism from these different standpoints should enable the identification of opportunities and difficulties which will suggest whether or not a new tourism policy is needed (step 3), or whether the existing development strategy (step 4), or action programmes (step 5) need be modified.

Acerenza (p. 60) defines tourism policy as 'the complex of tourism related decisions which, integrated harmoniously with the national policy for development, determines the orientation of the sector, and the action to be taken'. As such he sees tourism policy as providing the broad guidelines which shape the development of the sector, while the development strategy constitutes the means by which resources are used to meet the objectives defined. Acerenza suggests that three fundamental elements underlie all tourism policy: visitor satisfaction, environmental protection and adequate rewards for developers and investors. In this he appears to draw heavily on the earlier work of Gunn (1979, pp. 191–4). Acerenza, for example, stresses the voluntary nature of tourism while Gunn argues (p. 191) that: 'Tourism begins with the desires of travellers to travel and ends with their satisfactions derived from such travel and that planners should not lose sight of tourists' needs and wants' (Ch. 4).

If the previous stages have revealed the need for a new tourism policy then other forms of tourist development need to be explored and new goals set in the light of available resources and opportunities. Acerenza argues that such goals should reflect not only national priorities but also ensure the survival of the tourism sector itself. This might seem to be taking a very subjective pro-tourism stance but his argument is that only a stable, well established tourist industry can contribute to national or regional development. Developing a tourism policy requires both a clarification of basic goals and ensuring the compatibility of these for maximization of one goal might be at the expense of another. Gunn (1979, p. 194) distinguishes between goals which are continuous and 'are always being reached for' and more specific objectives which assist the transition from policy to action.

With stage 4, the determination of the development strategy, the emphasis shifts to the means of developing available resources to meet the goals outlined in the tourism policy. A first step here is a matching of supply and demand, an evaluation of tourist resources and markets using techniques outlined in Chs. 4 and 5. Attention must also be paid to competition for the products and markets identified and to the economic, social and technological resources needed to develop them. Alternative development strategies can then be assessed in the light of the goals and objectives set earlier and operational considerations to follow so that a final development strategy with specific objectives may be determined.

Stage 5 completes the transition to the operational phase where specific action programmes are defined in five main fields: institutional organization, development, marketing and promotion, manpower development and financing. Finally, Acerenza suggests a feedback loop from this stage to the initial ones whereby the results of the action programmes are fed into the analysis of previous development. Tourism planning is thus a long-term continuing process, a point also stressed by Baud-Bovy (1985) and other recent writers.

Figure 7.1 is a useful summary of recent trends in the literature on tourism planning, highlighting a movement away from an earlier emphasis on physical planning limited essentially to stages 4 and 5 to a more comprehensive approach incorporating a broader range of issues and actors (stages 1 to 3). However, many elements of these earlier stages are as yet little more than ideals which have rarely been formally included in tourism planning, whether through technical deficiencies, ignorance, lack of will or other real world limitations. Subsequent sections of this chapter thus deal essentially with the latter stages of Fig. 7.1 as aspects of tourism plans at different scales are discussed. Finally, constraints on planning are considered and conclusions drawn.

INTRAREGIONAL PLANNING

Intraregional planning, to use the WTO term for planning involving two or more countries from the same region, has attracted little attention in the literature. However, as the WTO's (1980c) inventory shows, intraregional planning is carried out in many parts of the world. The main emphasis at this scale is on joint marketing and promotional plans, with countries joining together to project a stronger regional image or undertake market research which otherwise would be inadequate or ineffective if done by individual States, especially smaller ones such as those in the South Pacific. An intraregional approach can be especially important where the product promoted appeals to international circuit travellers or wanderlust tourists. South American countries have jointly marketed the Andes Route while Honduras, Guatemala and Mexico have promoted a Maya Route.

Joint developmental and infrastructural planning for tourism is less common, perhaps because the associated investment is more tangible and located in specific countries. An example here is a joint plan by Argentina, Brazil and Paraguay for the development of the tourist potential of the Iguazu Falls, an important natural resource which straddles the three countries. Tourism is also an element of the Blue Plan, a large-scale intraregional planning programme involving seventeen countries which was set up by the United Nations Environment Programme to safeguard the environment and promote the harmonious development of the Mediterranean.

PLANNING AT THE NATIONAL LEVEL

As Acerenza (1985) and others have argued, planning for tourism at the national level, as at other scales, should be undertaken in the light of broader national development goals and objectives. In some cases these may be stated explicitly or there may be firm government direction. Elsewhere they may not be clearly articulated and the tourism planner has little guidance as to what overall goals should be pursued through the development of tourism. In many instances tourism plans focus specifically on tourism goals with little direct reference to broader issues, thus lessening the likelihood that tourism will contribute effectively to national development.

Earlier chapters have shown tourism is frequently seen to further national development by improving the balance of payments situation, by generating employment, spreading growth and so on. Where tourism is related to broader goals this is often done in economic terms, although growing emphasis is placed on social and environmental goals. Planners in West Germany (Klöpper, 1976; Romsa, 1981) and Switzerland (Keller, 1976), for instance, stress the need to provide recreational facilities for their urban populations and to protect the environment.

Ashworth and Bergsma (1987, p. 154) note that recent Dutch policy has stressed the need to mitigate the large adverse balance of international tourist payments (Table 6.2) by the stimulation of four aspects of the tourist industry:

- The capturing of a larger share of the high-spending intercontinental tourist trade to Western Europe . . .

– The diversion of part of the Dutch market from foreign to domestic
destinations, as a form of 'important substitution'
– The encouragement of 'near-neighbour holiday-making' especially from
the West German market, in order to exploit trends either to more
off-season holidays or summer holidays closer to home
– . . . more profitable exploitation of European transit tourists.'

Cyprus tourism policy aims to attract tourists from the high and
middle income groups (Andronicou, 1983). Not only are the economic
benefits from these groups perceived to be greater but a low-volume
high-spending market is believed to be more in keeping with a small
island where (p. 210) 'mass tourism would have had adverse effects
both on the environment and the social fabric of the country with a
consequential deterioration of all tourist attractions'.

Malaysia is one country where tourism planning has been tied
directly to broader development goals (Din, 1982). The 1975 tourism
development plan was formulated, *inter alia* '. . . to provide a basis upon
which Malaysia may develop her tourist potentials in an orderly and
balanced manner within the framework of the national development
plan and the New Economic Policy'. The New Economic Policy which
emerged during the 1970s had national unity as its stated primary
goal, national unity being sought by eradicating poverty irrespective of
race and by eliminating racial and spatial imbalances in the economy.
Thus the 1975 master tourism plan attempts to decentralize tourism
from the urban areas of the west coast of Peninsular Malaysia by
developing regions and tourist corridors on the east coast (Table 3.4)
and in the States of Sabah and Sarawak. While this may reduce some
of the spatial and associated regional imbalances, Din points out that
the plan makes no specific suggestions as to how local or *bumiputra*
(indigenous group) participation in tourism might be encouraged. He
also notes the practical difficulties of reconciling the double tasks of
promotion of growth, where the existing pattern of demand is centred
on the established tourist centres of Kuala Lumpur and Penang, and
the promotion of redistribution.

In his study of the Cook Islands, Milne (1987) highlights the
tension which may exist between different development goals and the
role of different strategies in attaining them. In Fig. 7.2 the two major
objectives of the Cook Islands government with regard to tourism are
shown to be maximizing gross tourist revenue generation on the one
hand and maximizing local participation on the other. Intensification
of the present pattern of development characterized by large foreign
or European-owned establishments with minimal local linkages might
enable the first objective to be met but local control would be sacrificed.
Conversely, encouragement of alternative tourist development would
enhance local participation and reduce leakages but at the expense of

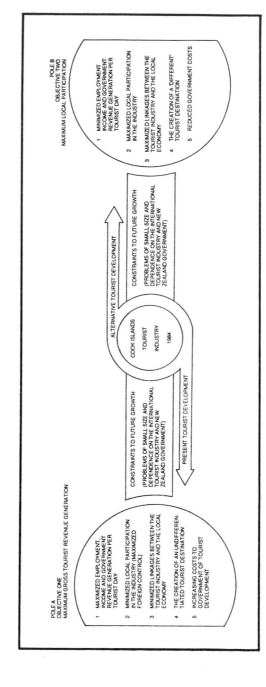

Fig. 7.2 Cook Islands tourist industry development continuum.
(*Source:* redrawn from Milne, 1987).

lower tourist expenditure per day. While not intending to prescribe a strategy for the expansion of tourism in the Cook Islands, Milne suggests that an optimal solution for future development continues to lie in a path between the two poles.

The total or partial incompatibility of different objectives was also recognized explicitly by the British Tourism Association (BTA, 1976) in its 1976 strategic plan for overseas tourism. The BTA ranked its primary objectives (p. 2): 'in accordance with current government policy of giving preference to the earning of foreign exchange in cases where the regional spread of traffic might compromise such earnings', and considered the interaction among various operational objectives which might achieve these (Table 7.1). Reconciliation of different tourist development goals may clearly be no easy matter but the task is facilitated if reference is made to broader national goals and the nature of the relationships between different objectives is examined at the outset.

Even where regional redistribution is not a primary aim, spatial considerations arise for planning cannot be done in a vacuum and national or regional goals must be translated into geographic terms.

Table 7.1 Possible tourism objectives

Primary objectives	Possible operational objectives				
	Maximize in total	visitor per visit	revenue per day	Maximize regional spread	Maximize seasonal spread
Improving foreign exchange earnings/ balance of payments	Y	(Y)	X	X	X
Raising incomes	Y	(Y)	X	X	(Y)
Redistributing incomes	X	X	X	Y	X
Maintaining employment	Y	(Y)	X	(Y)	XY
Conserving environment and heritage	XY	XY	XY	XY	(Y)
Fostering the arts, amenities and services	XY	XY	XY	XY	XY
Trade and goodwill	(Y)	X	X	(Y)	(Y)

(*Source*: BTA, 1976).
Y = Operational makes major contribution toward primary objective.
(Y) = Operational makes some contribution toward primary objective.
XY = Operational partly contributes to primary objective but is partly at variance with it.
X = Operational may be at variance with primary objective.

Commenting on tourism planning in Turkey, Ersek and Düzgunoglu (1976, p. 69) note:

> Significant problems in this field have been encountered during the implementation stages of the First Five-Year Plan and during the first two years of the Second Plan (1963–72) due to lack of policy decisions and tools to indicate the spatial distribution of resources and priority areas.
> The definition of priority areas and policy decisions concerning the geographical distribution of resources is of prime importance for a country like Turkey. This is especially so because of the scale and variety of Turkey's tourism resources.

A major concern at the national level therefore is to determine the most important regions to develop (United Nations, 1970). For those nations with an as yet poorly developed tourist industry, this will involve an examination of the country's tourist resources along the lines discussed in Ch. 5, together with an assessment of likely demand (Ch. 4), and the delimitation of one or several areas to develop. In other nations where tourism is already a significant activity, resource evaluation and marketing techniques may need to be applied to the question of where or how to concentrate future growth. Elsewhere, the prime concern may be to identify and rectify bottle-necks and deficiencies in the national tourist system rather than to promote new areas. This is especially important in countries such as New Zealand where the emphasis for overseas visitors at least, is on touring rather than destination-oriented tourism (Forer and Pearce, 1984). Bottle-necks in one or two key places may effectively limit the growth of tourism throughout the country as a whole.

Where possible, selection of areas should be guided by national planning objectives. In Mexico, for example, specific planning criteria have been defined (Collins, 1979, p. 354): 'New tourist centres should develop new sources of employment in areas with tourist potential. These areas should be located near important rural centres with low incomes and few alternatives to develop other productive activities in the near future. New resorts should spur regional development with new agricultural, industrial and handicraft activities in the zones.' These and other factors led to the development of Mexico's first planned tourism complex in economically depressed Quintana Roo (Cancun) in preference to other more developed areas such as Acapulco, or equally depressed regions like the Coast of Oaxaca which, however, lack infrastructure.

One of the objectives of Thai national planning has been to foster growth selectively throughout the country by designating key development areas. Elsewhere, for example in Bali (Rodenburg, 1980) and Tanzania (Jenkins, 1982), development has been geographically concentrated to limit some of the socially disruptive effects of tourism.

Spatial co-ordination with other sectors of the economy is also important at the national level. Development of major infrastructure such as roads, airports or ports, for example, should take into account not only the needs and demands of tourism but also those of other sectors such as agriculture or manufacturing. In some instances, multiple use of new infrastructure has been severely limited as Rodenburg (1980, p. 183) notes in Nusa Dua, Bali: 'Only $8.2 million (22.6%) of the planned infrastructure budget of $36.1 million will be used for multi-purpose roads and is the only item, with the exception of water and electrical supply lines to two local villages, that can be said to serve both the Balinese and the tourist industry.' Co-ordination is also necessary to ensure that valuable tourist resources are not destroyed by other activities. Ciaccio (1975), for instance, observes how the tourist development of much of the Sicilian coastline was seriously compromised by the installation of large oil refineries.

Structurally, one of the major concerns at a national level will be linking development areas with gateway cities. In many countries the majority of arrivals by air will be to a single city and links with the hinterland, particularly in developing countries, may not be strong. Even in economically advanced countries where such linkages exist, for example, Japan and the United Kingdom, international tourists may still be heavily concentrated in the capital city (Pearce, 1987a). Selective development of a small number of key tourist regions will reduce the number of linkages to be developed and at the same time permit the promotion of a stronger, more coherent image of these regions enabling them to compete more effectively with the points of arrival. However, in large countries where the potential is for touring rather than resort development, promotion of a second major entry/exit point will help increase the flow of visitors throughout the country and obviate the need for them to backtrack. But where the concern is for social tourism and promoting the recreational opportunities of nationals, the emphasis will be towards developing more localized linkages between the major urban areas and their immediate hinterlands.

In most cases, development of domestic tourism and international tourism will be complementary. The Dubrovnik Seminar (United Nations, 1970) noted:

> ... in some countries development of domestic tourism might lead to a development of foreign tourism, whilst in others, as yet undeveloped, but well endowed with tourist attractions, the encouragement of foreign tourism would lead in due course to growth in domestic tourism. In both cases, however, tourist development plans should provide from the start for both foreign and domestic tourists.

Many of these spatial elements were incorporated in a recent 10-year development plan for tourism in Belize (Pearce, 1984).

There, the resource evaluation, the analysis of existing demand and the marketing study suggested a developmental strategy focused on the main gateways, the northern cays and the southern coastal zone. Emphasis was given to facilitating access from the gateways to these coastal areas and to improving the distribution of information at the points of arrival in order to encourage longer visits amongst overland travellers by directing them out to the more attractive cays and along the coast. In these latter areas a prime concern was to upgrade the infrastructure.

Two more detailed examples of planning for tourism at a national level will now be given. The first, drawn from Vanuatu, involves a general national development plan with a tourism chapter. The second, the Thai national tourism development plan, provides a good example of the way in which many of the elements discussed above are brought together.

General National Plan – Vanuatu

The small South Pacific nation of Vanuatu (1981 population, 120,000) became independent in 1980. Like a number of other newly independent developing countries, Vanuatu sought to guide its future development through the formulation of a national development plan. The country's First National Development Plan covers the years 1982–86, with one of its twenty-seven chapters being devoted to tourism (National Planning Office, 1983). The reasonably cautious attitude to tourism expressed in that chapter and reflected by subsequent government activity must be seen in the light of the Plan's general objectives, including those to:

- preserve the cultural and environmental heritage of the nation through balanced programmes of development,
- ensure the progressive achievement of the long-term goal of national self-reliance.

In particular, the Vanuatu Government is seeking to develop a tourism sector with increased levels of local inputs, participation and control and one which is compatible with the nation's culture and environment. The government itself has taken a lead in some of these areas, forming a national airline, Air Vanuatu, in association with Ansett Airlines of Australia and establishing Tour Vanuatu as the dominant local tour operator. However, it has not intervened directly in the accommodation sector where lack of capacity is seen as a constraint and where externally financed hotels are recognized as being 'not necessarily a bad thing' (National Planning and Statistics Office, 1984, p. 158). Such hotels would be restricted to the islands of Efate (notably to Port Vila, where most accommodation is now located), Tanna and Espiritu

Santo with accommodation elsewhere being 'reserved for ni-Vanuatu projects of a small (bungalow) scale and then only if requested by the local community' (National Planning and Statistics Office, p. 167).

National Tourism Development Plan – Thailand

The development of Thailand has been guided by a series of 5-year national economic and social development plans since 1961. Little emphasis appears to have been given to tourism in the first three of these covering the years 1961–76 (Cavallaro, 1985). However, following the rapid growth in international tourism over this period which saw the sector become the country's third largest source of overseas funds, the Tourist Organization of Thailand called for the preparation of a more specific National Plan on Tourism Development. No reference to the broader national plans is made in the tourism plan which was to guide both government and private enterprise in meeting future growth by serving as a framework for feasibility studies and the development of master plans and detailed plans for selected tourism areas. The study was undertaken by an overseas consulting firm and a Thai one. The final report, on which this section is based, was published in 1976 (TDC–SGV, 1976).

Figure 7.3 outlines the basic study procedure which might be

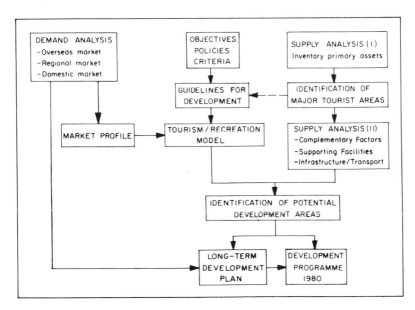

Fig. 7.3 Study procedure for Thai National Plan on Tourist Development. (*Source:* TDC–SGV, 1976).

considered a very useful general framework for tourism planning. Analysis of demand and supply was accompanied by the formulation of objectives and guidelines for development. This in turn led to the identification of potential development areas, the formulation of a long-term development plan and the establishment of a development programme to 1980. Market analysis suggested a strong growth of international tourism which would give from 2 to 2.5 million tourists by 1980. The majority of visits, given the short length of stay (average duration of 5 days by 1980), would be spent in and around Bangkok. However, the demand for touring and cultural tourism would grow, necessitating the development of regional and district centres and an increase in leisure-oriented international travellers would give rise to resort development. A strong growth in domestic tourism was also foreseen, but, given income levels, this would mainly involve day-trips and, to a lesser extent, weekend stays.

In terms of policies and guidelines, it was felt that product development should be market-oriented and the emphasis should be on unique Thai assets. Moreover (TDC–SGV, 1976, p. 3):

> Tourism facilities should be set up in such a way that they are suitable for integrated use by international and domestic tourists and recreation seekers of high, medium and low income structure. To make the investments economically viable and to prevent disturbance of the traditional way of living, dispersed concentration of the facilities is imperative. Infrastructure and superstructure should be as much multipurpose as possible, serving also other sectors of the economy. Existing plant should be fully exploited and upgraded before new development is started. Strict regulations should control the environment.

The analysis of supply began with an inventory of the country's primary attractions. Some 510 attractions were identified, classified (by the type and degree of importance) and plotted on a map. Subsequently sixteen major tourism areas were established using the following two major criteria:

1. Attractions have to lie within fairly short distances of each other.
2. These attractions need to have a high degree of quality and uniqueness.

These sixteen areas were then weighted by other criteria to establish development priorities (Table 7.2). In particular, with respect to the national tourism/recreation model, importance was attached to accessibility and pressure from urban areas.

Within these sixteen areas, a hierarchy of twenty-one tourist centres was then selected for long-term development. Three were to be developed as regional centres (first echelon), ten as district centres (six, second echelon; four, third echelon), and eight as resorts (fourth

Table 7.2 Evaluation of potential tourist development areas in Thailand.

(1) Major tourism areas	(2) Primary attractions	(3) Complementary factors	(4) Supporting facilities	(5) Accessibility*	(6) Urban recreation pressure	(7) Tourism pressure	(8) Score
Bangkok/Pattaya	+++	+	+++	++++	+++	+++	17
Chiang Mai	++	+++	+++	++++	+++	+++	14
Songkhla/Hat Yai	++	+++	++	++++	++	++	13
Phuket	+	+++	+	++	+	++	10
Hua Hin (Phetchaburi-Prachuap)	+	+	+	+	+	+	6
Kanchanaburi	+	++	0	+	+	+	6
Upper Central Region (Phitsanulok area)	+	+	0	++	+	+++	6
Pattani/Narathiwat	+	+	0	++	+	+	6
Khorat	+	0	+++	+	+	0	5
Ubon	+	0	+	++	0	-	4
Chanthaburi/Trat	+	+	0	0	+	+	4
Nakhon Si Thammarat	+	+	-	0	+	0	4
Chumphon	+	++	-	+	+	0	3
Trang/Phatthalung	++	++	0	+	0	0	3
Chiang Rai	++	+	0	0	0	0	2
Sakon Nakhon	+	+	0	0	0	-	1

(*Source: National Plan on Tourist Development*).

* For column 5 only:
all modes of transport incl. international airport
all modes of transport incl. semi-international airport
domestic airport
railway access
highway access only.

Score	
++++	4
+++	3	very good/high
++	2	good/high
+	1	fairly/moderate
0	0	poor/low
-	-1	very poor/low

echelon). In addition, twelve major towns with over 20,000 inhabitants were designated as tourism support centres in the overall tourism network (three, fourth echelon; nine, fifth echelon), only three of these being within major tourism areas. In the plan, Bangkok/Pattaya continues to be dominant but the other two regional centres of Chiang Mai and Songkhla/Hat Yai are to become the nuclei for tourist development in the North and South. Overseas tourists were expected to travel to these centres predominantly by air whereas domestic tourists would travel mainly overland. For the latter, intermediate centres would be important (district centres, country resorts and major towns).

The short-term programme (to 1980) aimed therefore at immediate development of regional centres in the most attractive regions and of major district centres located within the network of regional centres (Fig. 7.4). Specific development measures for these were then outlined. The plan also recommended the strengthening of the Tourist Organization of Thailand by its conversion into a ministry which would co-ordinate the various government and semi-government organizations and guide and control the activities of the private sector. Major projects would be carried out by special agencies in which the private sector and the government would co-operate under the supervision of a Tourism Development Authority. Minor developments would be implemented at a lower level. Manpower requirements were also considered and the government was urged to provide more training courses at different levels.

This encouragement of tourism in regional centres also appears to be reinforced by the country's fourth and fifth economic and social plans (1977–86) which aim, *inter alia*, at a more balanced distribution of the national economy and accord tourism more prominence (Cavallaro, 1985, National Economic and Social Development Board, 1987). It is also a policy which appears to have met with some success, for in 1986 Bangkok accounted for only 28 per cent of Thailand's accommodation capacity (Tourism Authority of Thailand, 1987). In absolute terms the forecasts contained in the tourism plan have also been reasonably accurate, with 2 million arrivals being recorded in 1981 and 2.8 million being registered in 1986. By 1983, tourism had surpassed both rice and tapioca as Thailand's single largest source of foreign exchange.

Nevertheless, a number of continuing problems and difficulties are recognized in the tourism chapter of the draft Sixth National Economic and Social Development Plan (1987–91). These include both external factors (worldwide economic recession, new competing destinations, negative images of Thailand abroad) and internal ones (problems in maintaining and developing attractions and services, difficulties with promotion and limitations on the scope of the Tourism Authority of Thailand). These and other factors indicate a trend towards a slowing down in the growth of international arrivals and an increase in domestic

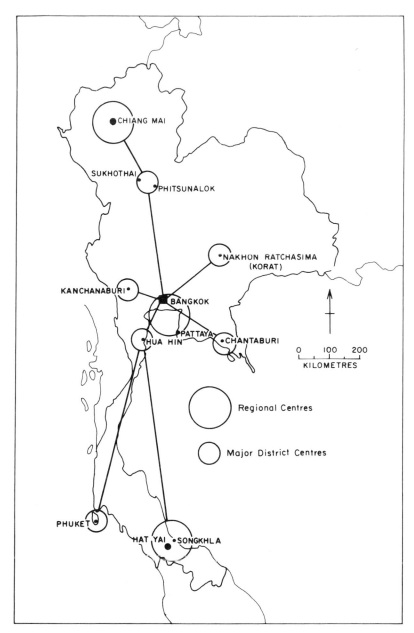

Fig. 7.4 Recommended centres for short-term development in Thailand
(1975–80).
(*Source:* TDC–SGV, 1976).

demand. The primary objectives for tourism in the new plan are:

– to achieve an increase in tourism, with particular importance being attached to employment and the local distribution of tourism benefits,
– to use tourism as a means of creating pride in being Thai.

More specific targets for increases in international and domestic tourism are then outlined, together with more detailed policies and strategies for the development of different tourism sectors.

PLANNING AT THE REGIONAL LEVEL

Tourist development regions will, in many cases, be defined by national plans although in other cases the initiative may come from the region. Tourist regions identified at the national level will usually be defined in terms of the spatial association of attractions and associated facilities, as in the Thai case (see also Ch. 5), or possibly in physical terms (a stretch of coast, a river system or a highland massif) or administrative ones, especially where tourism forms part of an overall regional strategy.

A first concern is with the choice of localities within the region which are to be developed. Some localities may have already been identified in the initial selection process but frequently more specific site evaluation will be required using the methods discussed in Ch. 3. At the same time, the regional objectives must be kept in mind, together with structural considerations. Where the aim is to limit social impact only one or two centres may be developed – the 'dispersed concentration' of the Thai plan – but where growth is to be spread throughout the region a larger number of development centres may be established. Economic and social objectives may also determine the choice between developing new functional resorts *ex nihilo* in virgin areas or grafting new tourist activities on to the existing settlement pattern, though this choice will also depend of course on the extent of existing development. In general, the degree of local participation and regional stimulus will be increased where existing settlements are incorporated into the development plan (see Ch. 3).

Some form of structural hierarchy will usually be necessary. A major regional centre may serve as a gateway to the region, provide many of the higher order services and functions, project a stronger promotional image and generally act as a development pole. Such a centre might be developed around the greatest concentration of attractions or a major settlement but some completely new base resort might also be created, as with La Plagne in the La Grande Plagne complex. Although a range of facilities might be offered in the regional centre, smaller centres might specialize in providing particular services or serving specific sectors of the market (e.g. families, the elderly, the sports-minded). The staged development of these centres may also allow for a more ready adaptation to changes in market demand. The transport network, which plays the essential role of linking these various centres together, may also be hierarchical.

Environmental considerations become important at the regional level where a range of spatial strategies may be implemented. Zoning measures may encourage the concentration or dispersion of tourist activity. The concentration option favours the location of all or most facilities in certain designated localities, preferably highly resistant environments as determined by studies of carrying capacity. Conversely, the dispersion policy encourages the distribution of smaller scale developments throughout the region, so as to reduce the environmental pressures on any particular spot. Concentration has been favoured in some coastal regions where a prime objective has been to avoid ribbon development the complete length of the coastline. In virgin areas such a policy may be reinforced by economic considerations; for example, the costs of providing infrastructure. A major concern in alpine areas is the altitude at which development should occur, with some writers favouring a move away from high altitude integrated resorts to development of accommodation at middle altitudes and linkages with the ski-fields (Chappis, 1974; Stanev, 1976). Areas to be left undeveloped, especially fragile areas, will need to be formally reserved by their designation as national or State parks, as some form of scenic or natural reserve, or by building codes. Dispersion may be more applicable to rural areas where facilities tend to be on a smaller scale. However, the uncontrolled dispersion of second homes can have unsightly results (Krippendorf, 1977).

A second technique is to relieve pressure on fragile areas by encouraging development elsewhere or by redirecting the tourist traffic. The impact on the coast may be reduced by measures to develop tourism in depth, that is, by distributing growth further inland. This may also spread the economic benefits. Or, access may be given to another forest or a second ski-field may be developed to reduce demand on areas reaching saturation. This is the so-called 'honey-pot' strategy which may be particularly effective if the new attraction intercepts traffic heading from the city to a pressure point. Krippendorf (1977) also suggests improving conditions within the city, so as to reduce the need to escape the urban environment, thus reducing demand on surrounding areas at its very source.

Several of these strategies were incorporated in the French regional master plans for the development of the coasts of Languedoc–Roussillon (Ch. 3) and Aquitaine which were prepared and implemented in the 1960s. Both of these were large-scale regional tourist development plans focusing on physical development and featuring a high degree of central government involvement in their conception and implementation, with major infrastructural work being undertaken by the State (Fig. 2.3).

The first plan for the development of the Languedoc–Roussillon littoral appeared in 1964, with modifications being made in 1969 and again in 1972. Although some of the details and names were changed in the process the original principles have been retained (Fig. 7.5). Firstly, it is a comprehensive plan which covers the entire length of the coastline in question and all facets of development. The plan's basic strategy was to concentrate development in five designated tourist units, leaving intermediate areas undeveloped. Such a strategy aims to avoid continuous ribbon development, to protect the more fragile parts of the coast, to bring certain economies of scale (an important consideration given the major infrastrucutral requirements) and to spread the economic impact evenly throughout the four departments of the region. Each unit incorporates a major new resort of 40,000–50,000 beds, which acts as a development pole, and several smaller existing ones which may be expanded or redeveloped. Development in the easternmost unit, for example, is centred on La Grande Motte. The traditional resorts of Palavas and Le Grau-du-Roi offered little possibility for further expansion but a major redevelopment project for Carnon was undertaken. Later, at the initiative of the Chamber of Commerce of Nîmes, a new marina was developed at Port Camargue. The Canet–Angeles unit is an exception, as development there has been based on the already established resorts of Canet Plage and St Cyprien. Incorporation of the old and the new broadens the market base, more readily ensures local participation while still permitting new concepts to be developed. Of the projected 400,000 new beds, 250,000 were to be built in the new resorts and 150,000 in the existing ones.

Although the coastline was divided up into tourist units, it was also necessary to retain some overall regional coherence. The plan provides for this by the motorway located some distance inland, to which the major resorts are linked by expressways. The motorway provides access throughout the region and a link with the rest of France. This is complemented by the chain of eighteen ports which assures a liaison by sea. Other major operations of a regional nature were those to eradicate the mosquitoes by drainage of marshes and extensive spraying, and to provide shelter belts and create wooded areas through large afforestation projects.

A similar approach was adopted for the development of the Aquitaine coastline, though the units there are not as large, in part a consequence of local environmental conditions. This is a sandy stretch of coastline, some 250 km long. The coast is backed by the extensive pine forest of the Landes, amidst which is located a chain of small lakes. Parts of this environment are fragile and unstable. There is but a thin soil layer

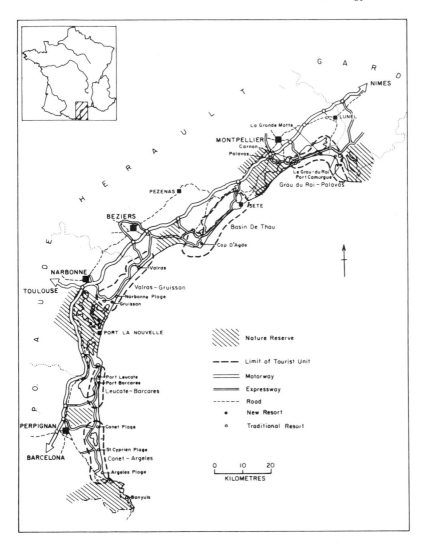

Fig. 7.5 Languedoc–Roussillon tourist development plan.
(*Source:* after Mission Interministerielle pour l'Aménagement
Touristique du Littoral Languedoc–Roussillon, Schema Directeur,
1972).

in the forest, thus the effects of trampling by large numbers of visitors
could be quite disastrous. The effects of trampling on the coastal dunes
could also be serious, for any blow-out would expose the forest to the
burning effects of the salt-laden sea breezes. Moreover the forests could

be suffocated by the invasion of the sand. The coast has been divided into seventeen sectors – nine development units (125,000 ha) and eight protected areas (265,000 ha). Again a regional motorway runs the length of the region, linking development units to each other and the region to the country as a whole. Smaller roads carry sightseeing traffic and cross protected areas. Each development unit is to be developed in depth. Such a policy is assisted by a more favourable microclimate in the forest and by the existence of an alternative to the coastal beaches in the form of small lakes located 10–20 km inland. This has enabled development perpendicular to the coast, whereby traditional coastal resorts have been complemented by small-scale subdivisions in the forest and other developments on the lake-shores.

Other regional tourism plans are less heavily oriented towards physical planning. In addition to some provision for new or upgraded infrastructure and plant such plans may also be concerned more with co-ordinating public and private sector initiatives and with marketing and promotion.

In Canada, planning for tourism at the provincial level has involved joint federal and provincial government action through a series of Travel Industry Development Subsidiary Agreements (TIDSA) to co-ordinate development and reduce fragmentation within the tourist industry (Montgomery and Murphy, 1983). The primary objectives vary from province to province, but a major concern is with upgrading facilities to enhance the area's attractiveness and to increase visitor numbers and their lengths of stay. Montgomery and Murphy provide a very useful evaluation of the TIDSA for British Columbia and while noting some success with the programme also identified five major weaknesses (pp. 202–6), namely:

1. The poorly ordered development process, with planning and strategy decisions following, rather than preceding, project funding.
2. The delay in processing grant applications.
3. Doubt about the public sector's role in tourist development.
4. Intra-group suspicion and rivalry within the private sector.
5. A lack of accountability in the development process and a lack of public input in the decision-making process.

In the absence of a national tourism plan in New Zealand, central government has recently encouraged regions, at the level of the united council, to take the initiative in tourism planning. Most effort appears to have been directed at regional promotion though a broader approach has been adopted in Northland where public input has been sought from the beginning and where the roles of the different regional bodies responsible for the implementation of different facets of the strategy have been recognized (Northland United Council, 1986). Figure 7.6 shows the regulatory and land use aspects of development planning are encompassed by the united council's regional planning scheme, the marketing plan is the responsibility of the regional tourism industry board, with the regional development council assisting with investment and information. Without some overall framework such as the regional tourism strategy these different functions can readily get out of step, while some important areas may get overlooked entirely or duplicated by different agencies.

Heeley (1981) draws attention to the multiplicity of inputs into planning for tourism at the regional level in Great Britain. In addition to the strategies for Britain's ten Economic Regional Planning Regions

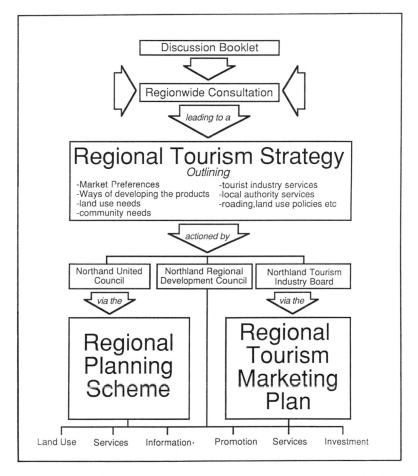

Fig. 7.6 Regional tourism planning process for Northland, New Zealand. (*Source:* after Northland United Council, 1986).

which generally have a tourism section, regional tourism strategies have been formulated by England's eleven regional tourist boards, regional strategies exist for sport and recreation, with a regional strategy for forest recreation also being drawn up for each of the Forestry Commission's conservancy areas.

Smith, Hetherington and Brumbaugh (1986) provide some useful practical insights into public participation in the preparation of a regional tourism marketing plan in Northeastern California during the early 1980s. The concept which emerged through expert assistance and community input was that of a tourism region united by Highway 89, a concept similar to other regional tourism plans based on theme itineraries or tourist corridors (Pearce, 1987a). Some success was achieved in developing this concept, in associated promotional activities and in attracting new touring business but the long-term effectiveness of the venture has been put in doubt by the failure to maintain a regional tourism council after an initial externally sourced establishment grant ended in 1984. In this respect, Smith, Hetherington and Brumbaugh (1986, p. 432) note:

> Provincialism and turfdom still hamper progress. Verbally the regional concept is vigorously approved but not reinforced with local money. Working together on a project funded by an outside force was acceptable but the concept of local joint funding is still not accepted.
> ... The parallel between rural tourism in northeastern California and Third World countries seems striking. Even when aided by expertise and socially nurtured, it requires the influence of outsiders, be they carriers, hotel chains or tour operators, to come, initiate and operate tourism enterprises.

The community approach was thus not without its limitations in this case but the authors nevertheless remain optimistic that the fostering of local leadership and the establishment of a broader information base will provide a firm foundation for future tourist development in the region.

PLANNING AT THE LOCAL LEVEL

Considerable variation exists in planning for tourism at the local level because of the immense variety of situations at this scale. A general concern is with land-use planning, with tourism constituting but one form of land use which the municipality or statutory local authority must plan for. In many instances, this is not positive development planning but rather is regulatory in nature and is concerned with reducing conflict over the use of land. Some local plans may be concerned with the management of specific areas, especially natural areas such as national parks (Heeley, 1981; Davies, 1987). In other instances, specific tourism development planning may occur as in the creation of new coastal or ski-resorts or in the development or conservation of urban areas. Many of these have been concerned primarily with physical planning and have not necessarily included marketing and promotional elements nor given much attention to the broader issues of development. Heeley (1986c, p. 29) observes that local planning authorities in Britain have traditionally been concerned with 'product and place, but not promotion and price'. A very useful review of tourism in structure plans in England is provided by White (1981). Specific tourism marketing and promotional plans are often prepared by local tourist boards having no statutory authority.

Whether the concern is with land-use zoning or development planning, a key consideration at the local level is the physical organization of the sectors discussed in Ch. 2. The primary attractions, be they natural or historical, will commonly be a focus for planning as their location is often fixed and their features vulnerable. The distinction between transport to and within the destination is critical at this level. The range of accommodation types, including residential housing, must be kept in mind and appropriate densities and height limitations determined. Provision must be made for services such as shops and restaurants as well as other functions and forms of land use, particularly where tourism is but one of several activities. Development will take several forms depending on whether planning is for a beach resort, an historic or cultural centre, or for tourism in an urban area. Details and emphasis will also vary from site to site but several general principles and considerations apply to most situations.

A first concern is not to compromise the site, either physically or visually, by the injudicious location of buildings or other facilities. Coastal sites may be particularly vulnerable. Development on or removal of the foredune, for example, can present serious erosion problems. Obtrusive sitings listed by the An Foras Forbartha (1970) report on amenity and tourism planning in Ireland include those on a ridge or hilltop, those breaking the waterline or not considering existing hedgerow patterns. Conversely, unobtrusive sitings include those against a backdrop of trees, hills or those taking advantage of existing landscape features. Other examples of integrative planning are given by the ICSID Interdesign Seminar (Gorman *et al.*, 1977) and Lawson and Baud-Bovy (1977). Coupled with this is the more general need to develop the area harmoniously. This implies an adequate balance between and within the different sectors in terms of capacity, quality and style as well as compatibility of different functions. Klöpper (1976), for example, notes the need for liveliness in some parts of spas and silence and cure facilities in others. Compatibility with non-tourist activities is also a major consideration. The design of the resort should be such that as many people as possible have ready access to the attractions and facilities. At the same time it must be recognized that different forms of accommodation are able to support different land prices or rents. In general, the more intensive the form, the higher the prices that can be paid. For instance, hotels will usually support higher prices than motels, and blocks of apartments more expensive sites than villas. Land prices will reflect the characteristics of the site and its location. Prime sites will be those with a better outlook and environment and will generally be located closer to the main attractions, though isolation may be a feature of some quality developments. Careful design, however, particularly of street lay-out, should still allow general freedom of access from most parts of the resort or centre to the main areas of recreational activity or tourist interest. Finally, the constraints of the site must be taken into account. In many coastal, alpine or other natural areas, only certain limited zones may be built upon due to soil or slope stability, the suitability of geological structures or drainage conditions. In historic or urban areas these constraints may arise more from the existing pattern of land use.

Several of these planning issues will now be examined in more detail with reference to coastal resorts (see also Pearce, 1987a). Aspects of planning in other areas have been touched on in earlier chapters while other writers provide useful reviews or examples of planning for other types of tourism at the local scale, namely ski-resorts (Sibley, 1982), urban areas (Chenery, 1979; ETB, 1981; Jansen-Verbeke, 1988) and rural areas (Shucksmith, 1983; Bouquet and Winter, 1987). Dean and Judd (1985) draw together a useful range of Australian examples.

Coastal resorts

Coastal resorts are perhaps the most common and distinctive form of tourist development, the coast being the premier tourist destination in many countries. Much of their distinctiveness arises from their location along the beach or seashore. For as Stansfield (1969) points out, growth outwards from a central core is limited to approximately only 180° as opposed to a full 360° in most other urban areas. Many of the older beach resorts, whether in England, France or the United States, have developed spontaneously along the traditional beach front or *front de mer* (Gilbert, 1939; Burnet, 1963; Barret, 1958; Stansfield, 1969). Typically, this consists of a parallel association of the beach, a promenade (or the boardwalk in the United States), a road or highway and a first line of accommodation and commerce where the best hotels and most expensive shops and apartments are to be found along with the casino. Beyond this, the intensity of accommodation decreases. This parallel structure now presents several disadvantages (Pearce, 1978d). Firstly, the first line of buildings, which are often high-rise in order to support the higher land prices, may constitute a barrier, both visual and real, between the interior residential zones and the beach or port. Secondly, the flow of pedestrians from these zones to the beach is disrupted by the automobile traffic of the intervening road, particularly if this happens to be a regional highway. Moreover, such a structure encourages linear or ribbon development, which is often not only aesthetically displeasing but also environmentally degrading, especially where continuous stretches of the coastline are developed. Resorts developed more recently in France exhibit a number of innovations and a greater degree of co-ordination. The solutions adopted by the newer resorts depend on their size and particular orientation.

The growing popularity of yachting and recreational boating in France since the 1960s has led to the development of specialized coastal resorts where the port is no longer a mere adjunct to the beach resort, but the very heart of the development. Port Grimaud, situated on the Mediterranean coastline in the Gulf of St Tropez, was developed by the architect F. Spoerry, whose basic idea was to bring together man, his home and his boat by the inter-penetration of the sea and dwellings. This was achieved by the development of small islands and interconnecting canals on formerly waste marshland (Fig. 7.7). The idea itself is not so new, for in many respects, including some of the architecture, Port Grimaud is a replica in miniature of Venice. What is innovative is the adaptation of this form for purely recreational purposes. The borders of each island have been developed as quaysides and mooring berths have been installed. As the dwellings themselves have been constructed virtually on the water's edge this enables each holidaymaker to literally step out the back door on to his boat or yacht. Essentially a parallel structure has been retained – canal, berth, footway, accommodation and vehicle access – but this has been so designed as to create a very functional resort with an accompanying holiday ambience throughout. The system of islands also assures a much greater length of mooring space, and limitation of the buildings to two or three storeys maintains a balance between the number of accommodation units and mooring berths. Height limitations were also imposed by the nature of the foundations and were aesthetically desirable because of the low-lying environment. A car park at the entrance to the resort receives the cars of visitors who must continue on foot, and those of the residents who generally have vehicle access only to deliver their bags and goods. Consequently a calm atmosphere prevails with most of the movement by foot or canal, either by private boat or by means of an electrically powered water bus.

The neighbouring Marines de Cogolin provides a different spatial expression of the same principle, the integration of the vacationer into his recreational environment, although the juxtaposition of accommodation and berths is less immediate than at Port Grimaud. The Marines de Cogolin is comprised of three main basins, with a smaller public port reserved for passing boats at the harbour entrance (Fig. 7.7). The accommodation has been developed around each of the basins, rising in a series of steps so that the maximum number of residents enjoy a view of the port and the activities that take place there. This arena-like

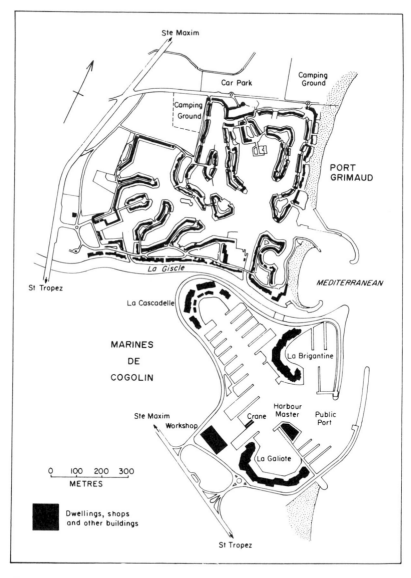

Fig. 7.7   Port Grimaud and the Marines de Cogolin.

architecture is also more aesthetically appealing than the typical vertical high-rise buildings which would jar in this low-lying environment. Most of the shops are located on the ground floor of the Galiote which also houses a cinema and night club. While temporary access to the quayside is permitted for loading and unloading equipment and stores, the general vehicle traffic is restricted to the far side of the accommodation where parking is available either in garages at the rear of the dwellings which protect the vehicles from the salt-laden sea air or in public car parks. Thus largely undisturbed access by foot between dwelling and boat is again possible.

Other comparable developments include Port Camargue, near the mouth of the Rhône, which, with a planned mooring space for 2,500 boats (one berth for less than five beds), is one of the largest recreational port complexes in Europe. Elsewhere smaller *cités lacustres* form part of a larger or existing resort. At Port Barcares, the Cité Nautica serves as a sub-centre, focusing activity on the Étang de Barcares, whereas most of the resort looks out to sea. It must be remembered, however, that all these developments are along the Mediterranean, that is, in comparatively low energy situations. The extent of tidal movements in the Atlantic has limited such projects along France's western coastline, without excluding them entirely; for example, Port Deauville. Moreover, concern has been expressed regarding the 'privatization' of the seashore, and private developments built on the immediate seashore have now been restricted. Nevertheless, many of the principles still remain valid elsewhere, particularly the attempt to separate the different forms of traffic.

In larger resorts having activities other than boating, the port may be only one centre of activity and the principle of a single row of dwellings surrounding the harbour or stretching out along the beach is no longer valid. This is especially true if ribbon development is to be avoided and the resort given some depth. The beach and port will remain key elements in the structure of any seaside resort but the traditional *front de mer* can be effectively modified as the new resorts along the Languedoc–Roussillon coastline have shown. La Grande Motte (43,000 beds) is one of the first and largest of these new resorts and embodies several new and interesting principles. A first concern at La Grande Motte has been to avoid some of the traffic and congestion problems inherent in the parallel structure of the traditional *front de mer*. The major access road still parallels the sea but at a distance of 200 m or so from the beach (Fig. 7.8). Secondary roads terminating in car parks run perpendicular to this main road giving access to the beach. Such a 'comb' or 'glove' system removes traffic from the beach area while leaving it readily accessible from all parts of the resort. To further encourage pedestrians, landscaped footways have been developed and no part of La Grande Motte is more than 10 to 15 min by foot from the

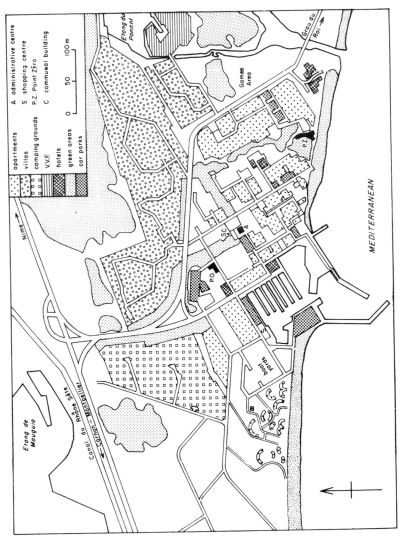

Fig. 7.8 La Grande Motte.

sea. As the densest accommodation is along the beach front, many of the residents live right on the beach itself. Point Zero provides a block of amenities for those using the eastern part of the beach. Overcrowding of the beach has been avoided as the amount of accommodation built has been a function of the beach's carrying capacity (see Ch. 5).

Although the average density at La Grande Motte is 100 beds per hectare, there is considerable variation in density within the resort itself (Fig. 7.8). The main access road cuts the resort in two. To the south, around the port and along the beach, are large, prestigious apartment buildings. Designed (by Jean Balladur) to ensure the maximum amount of insolation, these pyramid-shaped buildings give La Grande Motte its distinctive character and a general harmony of style although they have not been greeted with complete acceptance (Cazes, 1972). The ground floors of these pyramids are often occupied by shops and businesses, especially those around the port where many restaurants and terrace cafés are found. These multiple-unit dwellings become less dense and more conventional in form towards the interior of this zone, particularly to the east of the Avenue Melgueil where several buildings belonging to social organizations or housing more permanent workers are located. Few of the hotels have been able to locate on the most prestigious sites. The centre of the resort is comparatively empty, being occupied by gardens, a recreational arena and administrative buildings. Three distinct *quartiers* are found to the north of the access road. Here a zone of individual houses and villas separates the camping grounds in the north-west from a family holiday camp (VVF) (Village Vacances Familiales), located on the shores of a secondary activity zone, the Étang du Ponant, which offers a range of water sports. More recently the area to the west of the port has been developed. Known as La Motte du Couchant, buildings there take a crescent-shaped form and tend to be lower lying than in the rest of the resort.

In the cases mentioned above new ideas and principles have been able to be incorporated freely. However, the renewal or expansion of older beach resorts is constrained by existing developments. In most cases the prime beach front sites have already been occupied. Thus it may be essential not only to renew existing developments, but to stimulate and concentrate further growth around a new centre of activity. Construction of ports or marinas has been successfully employed in Languedoc–Roussillon to give new life to existing resorts. At Carnon and St Cyprien, for example, new apartment blocks have been built around newly created ports. Although deprived of a view of the beach, the residents of these apartments enjoy the comings and goings of the port and so have a sense of participation in the life of the resort. Furthermore, through traffic at Carnon is gradually being diverted to a new highway at the interior of the resort and the present road forming part of the *front de mer* will eventually be closed off to restore some tranquillity to the area. Access roads running perpendicular to the highway have opened a new zone of individual dwellings behind the band of existing beach front villas. As a result 7000 new beds will be made available, more than doubling the existing capacity of Carnon.

The Dutch resort of Scheveningen has also been successfully 'revitalised' through redevelopment based on a physical master plan, detailed marketing and a close association between the local authorities and the private sector (van der Weg, 1982; Chamberlain, 1983). The aim has been to broaden the traditional seaside resort product by developing a wider range of attractions to draw visitors the year round. Physical redevelopment of the seafront has been accompanied by the provision of new and varied recreational facilities, including an entertainment centre, circus, theatre, casino and speciality shopping centre. Promotional activities, a necessary adjunct, have been undertaken by a newly formed Scheveningen Resort Board which brings together representatives of the local authorities, the private sector and the local population. Visitor numbers and overnight stays have increased as a consequence and demand is now spread more evenly throughout the year. At the same time, van der Weg draws attention to the long-term nature of the redevelopment process versus the short-term policies of local authorities and to the inconvenience which redevelopment of an operating resort can bring.

Other problems have been experienced in older coastal resorts. In Brighton, Gray and Lowerson (1979) note that planning attempts have been ambivalent. Measures to control the traffic and parking in the town centre have been offset by the encouragement of new traffic generators such as the marina and the conference trade. Conflict has arisen over the desire to protect historical facades and areas of scenic beauty and the demand for the development of much-needed low-cost housing.

CONSTRAINTS AND CONCLUSIONS

The last two decades have seen a widespread, but by no means universal, recognition of the merits of planning for tourism as testified by the WTO's (1980c) inventory of tourism plans cited in the introduction and the other examples discussed throughout this chapter. However, even when plans are prepared they are not necessarily implemented successfully and effectively. Indeed, the WTO study revealed that of the more than 1600 plans inventoried, approximately one-third were not implemented. Moreover, many of the supposed desirable or critical elements of tourism plans reviewed earlier were found to be lacking in those the WTO analysed. In particular, few plans integrated tourism within broader socio-economic development objectives while (p. 22) 'tourism plans whose *social aspects* have priority over direct profitability are even more exceptional'. Few examples were also found of plans that made firm and specific provision for protecting the environment. Furthermore, most tourist development plans were not supported by any legislation thus obliging (p. 23) 'the responsible authorities to resort to existing legislation in different fields to enforce regulatory plans'. The WTO concluded (p. 22): '*a desire to plan* exists in the tourism sector, but . . . few countries have been in a position to follow a policy of continuity regarding tourism development. Furthermore, the virtual absence of legislation seems to prejudice applying a directive plan.'

In addition to inadequate data, insufficient funding or the lack of trained personnel to carry plans and projects through, problems which beset other sectors too, planning for tourism is complicated and constrained by its fragmented and composite nature and by variations in demands, functions and responsibilities from one scale to another. Ashworth and Bergsma (1987, p. 155) express the situation in the Netherlands thus:

> The nature of the tourist industry allows it to reflect, and in some ways even magnify, two fundamental difficulties. First there is a fragmentation of organisations for the formulation of public policy, and thus a proliferation of goals and instruments, operating at different spatial scales or on different aspects of this inevitably broad activity. This in turn has led as a reaction to a search for integration both functionally and administratively, through national and local policies. Secondly there is uncertainty over the balance between the roles of the public and market sectors of the economy in supplying different aspects of the tourist product, which in turn has led to the search for 'partnerships' in various ways, at various administrative levels. The respective role within such partnerships however, remains to be determined.

Similar concerns are echoed by Din (1982) in Malaysia and by Heeley (1981) in Britain. Heeley (p. 75) argues:

> . . . a more positive national approach is needed. The Government's present objectives in tourism amount to little more than a statement of good faith in the balance of payments and regional economic development contributions of tourism and in the desirability of spreading tourist flows in time and space. They do not provide a basis on which to implant a national perspective on the strategies for tourism which are emerging at a regional and local scale.
>
> A consistent theme of local scale issues and policies is the discouragement of self-catering forms of tourist accommodation such as caravans whose local economic contribution is relatively small. To what extent is there a need for a national policy to take account of this while at the same time establishing the desirability of cheap and acceptable holidays for the urban population?

Torres Bernier (1985) points to the new division of responsibilities which have resulted from the creation of autonomous regions in Spain during the early 1980s. Central government is left with the role of promoting tourism abroad, managing pre-existing facilities such as the chain of Paradores Nacionales as well as with more general policies affecting tourism, such as those of economic stabilization. The autonomous regional governments, such as those for Catalonia and Andalucia, now have responsibility for the development and planning of tourism plant, domestic tourism and related areas of interest, for example the environment, regional planning and leisure and recreation. While the full impact of these changes has yet to be seen, Torres Bernier suggests the relatively indiscriminate development of accommodation under national policies dominated by balance of payments considerations may no longer be tolerated by the regions.

In Yugoslavia Vukonic *et al.* (1978, p. 24) note that physical plans for tourism are only informal legal and technical documents whose implementation depends on smaller scale urban plans which are often changed by local authorities motivated more by short-term interests than long-term considerations.

Outright opposition to plans seeking to control the development of tourism may also reduce their effectiveness or lead them to be shelved entirely. Morris and Dickinson (1987) illustrate these issues graphically with respect to the introduction of new local physical plans along the Costa Brava:

> ... the opposition has often been virulent, stemming from a mixture of incomprehension of the nature of planning, a deep-rooted suspicion of authority, and some manipulation by the commercial organisations which fear negative effects. In sum, the strength of the opposition to planning means that restrictions on the growth in tourism-related land use are still slight and likely to remain so for the foreseeable future.

Local opposition to plans fostering the expansion of tourism may also be encountered, as was noted in Ch. 6 with regard to Corsica (Kofman, 1985; Richez, 1986).

Despite these constraints, shortcomings and difficulties, planning can make a positive contribution to tourist development as examples in this chapter – Thailand, Languedoc–Roussillon and Scheveningen – and in earlier chapters have shown. Planning at all levels can help to increase the economic, social and environmental benefits which tourism may bring and at the same time reduce many of the associated costs. This can more readily be achieved where a broader approach is adopted, where goals and objectives are clearly defined and related to local, regional and national needs, where sound resource evaluation is complemented by an analysis of tourist demand and marketing and where a legal and administrative framework allows plans to be not only devised but also implemented. Greater recognition of the real world problems which may be encountered along the way may enable planners, developers, administrators and host populations to overcome them and thereby to translate some of the more idealistic goals of tourism planning into reality.

Review and reappraisal

This book began with a discussion of the nature of tourism, a review of the development debate and an examination of models of tourist development. Issues and questions raised in these sections have been pursued, explored and expanded in subsequent chapters. In this final chapter, three significant and recurring themes resulting from the initial discussion are reviewed or reappraised. First of all, there is a review of the systematic approach to the study of tourist development employed here, an approach which flows from the composite nature of tourism and the various models reviewed in Ch. 1. Consideration is then given to development both as process and state, firstly by way of a comparison of processes of tourist development in developed and developing countries, secondly, by a reappraisal of tourism's contribution to development. The directions tourist development might take in the future are then explored.

A SYSTEMATIC APPROACH

Attention in Ch. 1 was drawn to the composite, multi-faceted nature of tourism. This has led to a systematic approach being adopted throughout the remainder of the book whereby the emphasis has been initially on disaggregating tourism into its component parts, then identifying the relationships between them before drawing them back together again. This approach has been employed at the level of the individual chapter and of the book as a whole. The way in which it has been done has also been influenced by the questions and issues raised in some of the theoretical discussion presented in the first chapter. Questions of the what, when, how and why of tourist development raised in conjunction with the models of van Doorn, Miossec, Butler, Gormsen and others (Figs. 1.2, 1.3, 1.4, 1.5) have also been examined in terms of the underlying origin–linkage–destination framework (Fig. 1.1).

Thus Ch. 2 involved a detailed study of the different agents and elements of development while Ch. 3 attempted to establish some general processes by their synthesis in a series of typologies of tourist development. The value of these typologies, it is suggested, lies not only in the specific classifications given but also in the criteria and frameworks they provide for examining tourist development elsewhere. In Ch. 4 a consideration of the relationships between demand and development was followed by an examination of the two basic components of the demand for tourism – motivation and the ability to travel. This revealed the diversity of tourist demand and led subsequently to a review of decision-making steps, wherein different factors become important at different stages, and an examination of market segmentation. Likewise in Ch. 5, the many and varied factors influencing the distribution of tourist facilities and site selection are outlined then brought together, firstly in a review of evaluation techniques and secondly in a discussion of ski-field location. General frameworks for assessing the impact of tourism were presented in the introduction to Ch. 6 (Table 6.1 and Fig. 6.1) and followed by an examination of the more specific economic, social and environmental impacts which tourism may have. In each case, an attempt was made to isolate the impacts of specific aspects and processes of tourist development while retaining some overall perspective through the frameworks used (Tables 6.10 and 6.11, and Fig. 6.2). Methodological limitations and data deficiencies still exist but such an approach, it is argued, encourages a more balanced and holistic assessment of tourism's impacts. Recognition of the composite nature of tourism and the multiplicity of players involved in its development was also shown to be critical in planning for tourism (Ch. 7).

Throughout these chapters emphasis has also been given to the structure of tourism expressed in terms of the origin–linkage–destination framework (Fig. 1.1). The focus has not been simply on development within destinations but also on other associated aspects occurring in the origins (or markets) or relating to the linkages between the two (transport, promotion, etc.). In Chs. 2 and 3, for example, attention was drawn to the activities which occur in each of these three areas and how this at times fostered the emergence of multinational enterprises or limited the ability of local developers. Chapters 4, 5 and 7 stressed that analyses of demand and resource evaluation were related, not independent exercises (Fig. 7.3), that conditions in the origin were as significant as destination features in explaining the level and type of tourist development. Demand for tourism, for example, was interpreted by several writers in terms of spatial inversions (Table 4.1), the nature and extent of these inversions accounting in part for the type of experience sought in the destination and the resultant impacts. In Ch. 6, impacts were seen to occur not only at the destinations but also in the origins, even if little research on the latter has yet occurred.

This is not to suggest that there is no place for the more traditional case study, an approach which has characterized much tourism research, for indeed many of the general points made in earlier chapters have often been derived from such studies or illustrated by them. Nevertheless, the value of case studies could frequently be greatly enhanced if they were set in a broader conceptual context and if a more systematic approach was adopted in carrying them out. Any general increase in our understanding of the nature of tourist development will only come about if research efforts are cumulative, if general issues are identified and pursued and explicit links are made between different areas of research and the methods employed in them. An attempt to illustrate further the utility of such an approach is made in the next two sections where development processes are reviewed and tourism's contribution to development is reappraised.

DEVELOPED AND DEVELOPING COUNTRIES: A REVIEW

While sets of common processes were identified in Ch. 3, considerable variation was shown in the types of tourist development occurring in developed countries and, to a lesser extent, in developing ones. Any attempt to compare the processes found in developed and developing countries must therefore acknowledge the internal diversity which occurs, whether between different forms and types of tourism in any one country, for example France or Mexico, or between countries, particularly those in the Third World. Contrasts are to be found, for instance, between the small Pacific and Caribbean micro-States on the one hand and larger more diverse nations such as Mexico, Morocco and Malaysia on the other. Generalizations regarding tourism in developed and developing countries must necessarily be limited but certain points can still be explored fruitfully.

One of the few explicit comparisons made between aspects of tourist development in developed and developing countries is that by Jenkins (1980, p. 27) who draws attention to the role of government intervention:

> The governments of developed and developing countries share many areas of responsibility. But in the developing countries the problem of resource

scarcity and consequently allocation is acute. Strong government control is necessary to prevent exploitation and obvious waste, and to ensure that the benefits from tourism are optimised. Tourism in developed countries can be regarded as a mainly social activity with economic consequences: in developing countries it is largely an economic activity with social consequences.

The examples in earlier chapters have shown that public sector intervention in developing countries largely involves central government which acts to regulate or encourage tourist developers as well as directly participating in development. In addition, central government agencies may serve a a conduit for international aid. While central government is also involved in developed countries (Fig. 2.3), regional government and especially local authorities tend to play a much greater role there than they do in the Third World as the various coastal, ski-field, rural and urban examples showed.

In terms of the private sector, a number of studies of tourism in developing countries reviewed earlier have highlighted the role of transnational developers and emphasized the degree of control exercised from abroad. The report of the Centre on Transnational Corporations (1982) indicated the ratio of TNC hotels was certainly much higher in developing countries. At the same time it must be remembered that in absolute terms, TNC hotels are also significant in developed countries (Table 2.3). In the case of Paris, international and domestic chains were shown to be a major force in recent hotel expansion. Domestic chains have also played a significant role in the development of Cancun.

Other comparisons can be made between the types and processes of tourist development in the two sets of countries. One of the key features identified with tourism in the Third World is enclavic development, the development of isolated, usually large-scale projects undertaken often with a considerable degree of external participation, whether by means of investment assistance from international agencies or active promotion by multinational developers, primarily for foreign visitors (Fig. 3.7). In physical and financial terms, however, such enclaves have their parallels in developed countries, especially France (Figs. 3.2 and 3.3). The integrated ski-resorts of the French Alps and the development of the new resorts of Languedoc–Roussillon have much in common with the sun–sand–sea enclaves of Bali, Cancun, the Caribbean or the South Pacific. The main reason for this is that the advantages of scale economies and the freedom to develop an isolated site apply as much to resorts created *ex nihilo* in developed countries as to those built in the Third World.

Nevertheless, differences do occur. The 'external' developers in the French case are essentially external to the region or community rather than to the country. International aid agencies scarcely play a role in enclavic development in developed countries though for tourism projects within the European Community assistance may be obtained from the European Regional Development Fund. The clientele in the French case remains overwhelmingly domestic and for this reason the social considerations of enclaves are much less significant there than in developing countries where contrasts between local residents and the predominantly foreign visitors can be very marked. Moreover, in France there is a large proportion of independent holidaymakers not channelled through tour operators or travel agencies.

Changes in the extent of local/external participation have often been not well documented but the evolution in the structure of control does not appear to have been constant nor unidirectional. In several instances the pattern observed approximates that proposed by Butler (1980), with growing external domination in the development and consolidation phases followed by increased local participation as some decline in demand sets in. In the case of Jamaica, for instance, disinvestment by foreign hotel operators during a downturn in arrivals in the 1970s caused the government to take over the management of a number of hotels while in Tobago increased national participation followed the expiry of tax holidays offered to foreign operators and changes in the Trinidadian economy. It could be argued in these cases that the foreign developers took their profits in the good times and then left or conversely that they provided the basis for national involvement. However, any risks they took in the pioneering stages were softened by the support offered by the host governments. In Cancun, domestic demand was very important in the first years of development though much subsequent growth was in international arrivals. Domestic developers and chains were also important there, though supported by international infrastructural investments. An increase in domestic demand met in the main by national or local developers has also been observed in Morocco along the lines suggested by Gormsen (1981).

In the case of the integrated French ski-complex of La Grande Plagne, local failure was followed by large external inputs which did enable, however, some local participation in the smaller satellite resorts. At Vars, external developers provided the catalyst for more extensive local participation while in Austria local initiatives and control appear to have always been a characteristic feature of much ski-field development. In any event, local participation does not necessarily mean a more satisfactory form of development, particularly where a drive for uncontrolled expansion comes from only a few dominant local entrepreneurs or sectors of the community as Morris and Dickinson (1987) showed in parts of the Costa Brava and Bromberger and Ravis-Giordani (1977) illustrated in certain communes of Provence.

Similarities and differences are also found in more modest types of tourist development in developed and developing countries. The 'alternative tourism' of developing countries has its counterpart in the rural tourism of Europe, particularly in farm tourism. Both involve small-scale developments closely linked into the existing community and way of life. These usually feature traditional dwellings or modifications to them. However, alternative tourism has been much more limited and less successful than the farm tourism of Europe. Various interrelated reasons account for this. Farm tourism in Europe is accessible to a large domestic market which can drive to individual farms and *gîtes*, whereas the alternative projects in the Third World are frequently handicapped by the costs of international travel in the first place and poor local connections within destinations in the second. Differences in culture and standards of living between the hosts and guests are less pronounced in Europe than in the Third World. A shared language greatly facilitates communication between host and guest and eases overall organization. Finally, farm tourism in Europe has been marketed much more effectively and has been developed with the help of a variety of advisory and support services. The most successful of the alternative tourism projects – that in the Lower Casamance – also shares some of these latter characteristics.

What distinguishes tourist development in many Third World countries from developed nations is the comparative lack of intermediate cases, particularly in the small insular Caribbean and South Pacific States which have constituted the focus of the literature of tourism in developing countries. Few examples are found there of tourism which falls between large-scale enclave developments and the small and limited alternative tourism projects. Such a gap can be attributed primarily to the small size of such States, their limited tourist potential and above all to the virtual absence of any domestic market (a function of the small surface area and population as well as standards of living). The absence of domestic demand not only limits the range of facilities which are sought but also restricts the active participation of nationals in the development process. Entry levels into domestic tourism are generally lower and experience acquired there can form a basis for subsequently catering to international visitors. In the larger but less frequently studied developing countries cited – Mexico, Morocco and Malaysia – there is evidence not only of a significant domestic tourist market but also a range of development types and processes, some of which have an important national component (Table 3.4). Aspects of tourism in some of these larger developing countries start to resemble closely those found in developed countries.

Overall, many of the general differences between tourist development in developed and developing countries appear to be differences in degree rather than in kind. Moreover, differences within countries, particularly larger ones, may be as significant as those between them. In this respect, the scale of analysis is important. Peripheral regions within

countries, for example parts of the French Alps, may exhibit characteristics of developing countries and the processes and consequences of
tourist development found there may be analogous to those identified in
the Third World. Indeed Dorfmann (1983) and Guérin (1984) present
arguments similar to those advanced by Hills and Lundgren (1977),
Britton (1982) and others for the Caribbean and South Pacific.

TOURISM AND DEVELOPMENT REAPPRAISED

Any general assessment of the contribution which tourist development
can make to the broader development of communities, regions and
countries based on analysis of the impacts examined in Ch. 6 is
conditioned by their diversity and complexity and by one's interpretation
of development as discussed in Ch. 1. The economic impact studies
cited continue to reflect a dominant view of development as economic
growth, with tourism's contribution to the state of development being
expressed in terms of foreign exchange earnings, GDP, employment,
State revenues and other such measures. Overall, the impact in these
terms has generally been shown to be reasonably positive although less
extravagant claims are now being made and a more balanced picture
has emerged as techniques have been refined and as associated costs
are more frequently being taken into account. For example, the
extremely favourable view of the foreign-exchange earning capacity of
international tourism which was widely held in the 1960s was sharply
tempered in the 1970s as associated leakages were acknowledged.
Subsequent research beyond the small developing countries where
much of this second stage research was initially undertaken has resulted
in a broader perspective and one which underlines the significance of
the type of tourist development and the economic context in which
it occurs as factors influencing tourism's contribution. Likewise,
Latimer's (1985) review of tourism and agriculture brings into question
some of the evidence which emerged in the previous decade regarding
the extent of inter-sectoral competition.

Further reassessment can be expected in the future as additional
research is undertaken in a wider range of contexts and as tourism and
other sectors of the economy continue to evolve. However, although
reassessing the evidence available from time to time is valuable, more
effort might also be directed at attempting to measure tourism's
contribution to development over a given period, that is, attempts must
be made to take account of both the process and state of development, as
for example in the longitudinal studies by Préau (1983) and Getz (1986a)
discussed in Ch. 6. Methodologically this is a very demanding task.
In terms of economic growth, econometric modelling appears to offer
some particularly useful insights, enabling the with/without tourism

elements of the framework presented as Table 6.1 to be incorporated in the assessment.

Courbis (1984) reports on a study involving the construction of a regional econometric model which enabled a simulation of the economy of Languedoc–Roussillon with and without the large-scale tourism project discussed in Chs. 2, 3 and 7. The results of this analysis showed that by 1980 the regional output was from 10 to 14 per cent greater than it would otherwise have been and that from 28,000 to 31,000 supplementary jobs had been created. Courbis considers that the creation of these jobs has been very cost effective. An investment of 1 million francs (at 1983 values) in tourism (including infrastructure and accommodation) was estimated to create 1.7 net jobs. In budgetary terms, a 1 million franc State subsidy led to the creation of from eleven to twelve net jobs which meant a cost to the State budget of from 80,000 to 90,000 francs per job. Unemployment for 1980 had been reduced by about 20 to 28 per cent. Although the region continues to record one of the highest rates of unemployment in France, there can be little doubt that without the development of coastal tourism the economy of Languedoc–Roussillon would be far less healthy than it is today.

A similar econometric modelling exercise was undertaken by Archer (1984) in Barbados. Calculations of changes in GDP first included then excluded the impact of tourism-related spending in the economy, the difference between the two results providing an estimate of the contribution made by tourism to the growth of the island's GDP. Archer (p. 11) concluded that:

> The tourist industry was indeed a successful growth-generator during the last two decades, even in the context of significant import leakages. For example, from 1961 to 1977, measured in constant dollars per capita income almost doubled and two-fifths of the increase was attributable to the growth of tourist income. It appears that the public investment in tourism was a wise one.

He then continues by asking:

> Is the wealth created by tourism widely shared or does it increase social and economic inequalities? Does tourism-based economic growth lead to structural transformation of the economy, or is 'growth' in this case different from development? Empirical research on these issues will do much to further clarify the true nature of the tourism impact currently being felt not just in Barbados but throughout the Caribbean.

These questions start to address some of the other developmental considerations discussed in Ch. 1, notably development as distributive justice and as socio-economic transformation. Far fewer of the studies have been framed specifically in these terms but such questions have been posed in some of the work on tourism in the Third World and in some of the regional development studies. Rajotte (1987, p. 84) notes with regard to tourism in Kenya:

> As the drive of the 1960s for simple economic growth has given way to a concern in the 1970s for equity and the distribution of benefits, so too has been increasing disquiet that tourism's benefits go more to investors, transnational corporations, managers and directors than to the many persons who try to derive their whole or partial living from work within the tourism industry. Further, tourism's benefit to the nation as a whole is doubtful.

Allcock's (1986) assessment of changes in Croatia, Dorfman's (1983) less optimistic interpretation of development in the European Alps and the rejection of official tourist development plans in Corsica (Kofman, 1985; Richez, 1986) exemplify the different situations which have been identified in the regional development literature. While considerable variation is to be found, the general situation appears to be one in which the economic benefits of tourism are neither as widely penetrating nor as narrowly restricted as some earlier writers have claimed. For some regional economies and developing countries the expansion of tourism has clearly been associated with a growing dependence on external developers and inputs, with some loss of control over the path development has taken and increase in leakages from the national and regional economies. Such a pattern, however, is by no means universal nor inevitable as the preceding section has shown.

In one of the earlier studies which attempted to examine some of these broader issues, Boissevain (1977) tried to establish the Maltese perception of development through an analysis of Malta's development plans and found (p. 527): 'development to Maltese planners means economic growth, increased social equality, greater national cohesion, improved quality of life, and finally and most important, greater independence'. Boissevain then attempted to assess how tourism has affected each of these goals and concludes (p. 535) that it has 'a beneficial impact on development in Malta [which] contrasts with the dire consequences reported, predicted or inferred by many anthropologists'. One of the reasons for this, he suggests, is that he has attempted to evaluate the impact of tourism in Maltese terms rather than in his own, an approach not commonly adopted. Moreover, Boissevain gives weight to the local context, identifying (p. 536) 'circumstances peculiar to Malta which have facilitated a smooth adjustment to international tourism'. Several years later, however, Oglethorpe (1984) is writing of a 'crisis of dependence' in Malta, with British tour operators dominating the local market.

The work of Boissevain and others reinforces the importance of context and process in assessing tourism's contribution to development. These factors are equally important in evaluating the externalities of development as the discussion of environmental and social/cultural impacts showed. The results of the OECD tourism and environment programme, for example, indicated adverse impacts were more likely to be experienced where tourism developed rapidly with little planning and inadequate infrastructure and a marked seasonal concentration of demand. Considerable variation was found in the social and cultural impacts which result from different kinds and levels of tourist development, but in general, tourism was viewed more positively in the more mature destinations.

As with the economic impacts, the more recent environmental and social/cultural impact studies tend to be more balanced than those undertaken earlier which were often rather extreme in their evaluation, particularly when stressing the adverse impacts which tourism may generate. The picture which now often emerges is one which not only acknowledges that economic benefits are counteracted to some extent by economic, social/cultural and environmental costs but also that they may be complemented by environmental and social/cultural gains. Moreover, these issues are closely interwoven. Reduction of adverse environmental effects may have short-term economic costs, through for example, the expense involved in upgrading infrastructure, but in the longer term there may be a significant economic gain as the area retains its image and maintains or expands its clientele.

Attitudinal surveys have produced varied results but generally tend to show the economic advantages of tourist development are perceived to outweigh many of these other considerations, with an overall positive response being recorded in many destinations (e.g. Rothman, 1978; Belisle and Hoy, 1980; Pearce and Cant, 1981; Garland and West, 1985). More work is needed, however, in assessing the overall impact of tourist development by weighing economic costs and benefits expressed essentially in quantitative terms against social and environmental gains and losses, many of which can only be expressed quantitatively.

Fuller assessment of tourism's contribution to development also requires a more thorough evaluation of tourism's performance against that of other sectors. The majority of studies cited earlier have dealt solely with the impact of tourist development and provided no touchstone against which tourism might be compared. Those studies which have had a comparative element, for example the discussion of multipliers by Hughes (1982) and Cooper and Pigram (1984), have generally been more restrained in their evaluation of tourism's contribution to development.

At the same time, more consideration needs to be given to what alternatives to tourism there are or were in the areas in question. Tourism will only be a comparatively better or worse development option if there is indeed a range of possible development paths open. While few areas will have no choice other than to exploit what resources they have by developing tourism, a number of the studies cited here, for example Duffield and Long's (1981) review of tourism in Scotland, suggest tourism has often become important where other development opportunities have been limited. This does not mean that tourist development is either a last resort or universal panacea, for to be successful a region must first have the resources which might be developed to meet an existing or potential demand. Planning can enhance the way in which these elements come together and influence the extent to which tourist development contributes to a broader process of development but planning too is not without its constraints and limitations.

FUTURE DIRECTIONS

Many of the issues and questions raised in this book will remain important or become even more critical in coming years as the continued expansion of tourism into the twenty-first century seems assured. Despite the checks in international arrivals in the early 1980s and some downturn in domestic tourism, by the middle of the decade significant growth in tourism was again being experienced in many parts of the world (Fig. 4.3), a demonstration of its resilience to major economic and other pressures and an expression of tourism's significance in modern society. The growth rates of the 1960s and early 1970s may not be repeated and major regional markets and destinations may see their importance change but significant worldwide increases in the total volume of tourism will continue into the foreseeable future. No attempt is made here to predict future levels of demand; rather a number of major trends are identified and their implications for future development discussed.

Tourism will continue to expand for two main reasons. Firstly, the demand for tourism will grow as any analysis of the factors discussed in Ch. 4 indicates. Secondly, despite reservations expressed by certain communities, regions or nations at present or in the past, the development of tourism will continue to be encouraged or pursued by both the public and private sectors as many of the motives discussed in Ch. 2 will become increasingly important in a post-industrial age. At the same time changes will occur, both in the nature of demand and of tourist development.

Major factors influencing the demand for tourism include the availability of free time and disposable income. The WTO (1983a) study indicates a continuation of the trend towards increased paid holidays (Table 4.2), with the potential for this in developing countries being particularly great. In many developed countries, increased free time will also come about through the demographic ageing of the population and consequently more time being spent in retirement. Trends in unemployment will also influence the amount of available free time. Both in this latter case and in that of developing countries, growth in free time will be tempered by the corresponding increases in disposable income. Regional variations in overall economic rates of growth will thus be important. Total demand is also affected by total population size with significant worldwide differences occurring. The ageing of many Western populations has been accompanied by stabilizing or even decreasing populations while those of many developing countries continue to experience relatively high rates of growth.

Different patterns of demand might thus be anticipated in the future. One of the more significant trends may be a marked increase in domestic tourism in some of the larger and better off developing countries, the initial stages of which have already been seen in places such as Mexico, Morocco and Malaysia. Such a trend will alter existing patterns of ownership, development and impact.

An increase in mature travellers will occur in developed countries as their populations age. Chew (1987, p. 84) suggests that for this group the 'holiday-taking habit has become high on their list of priorities and is no longer a luxury' while Frechtling (1987, pp. 108–9) notes with regard to the United States market:

> The retired do not take as many trips away from home as other age groups, but when they do, they tend to stay away 50% longer and travel 10% further than their younger counterparts. They are also more likely to travel to a foreign destination.
>
> The growth of this segment to the year 2000 should boost long-distance and duration travel, travel to foreign destinations, sales of package tours, common carrier travel, travel during non-summer seasons, and travel by recreational vehicle.

Frechtling also draws attention to growth in the ethnic, singles and business markets.

Further fragmentation of holidaytaking in developed countries is also likely to occur along the lines discussed in Ch. 4, with main holidays being progressively supplemented by shorter secondary ones. The WTO (1983e, p. 17) suggests individuals await these as 'brief periods of decompression' and see in them the 'most promising field for innovation, both quantitative and qualitative.' Innovations in this field include products offering a brief 'respite' or 'immersion' in a particular location or region, organized around a central theme such as the landscapes and customs of Scotland or the culinary delights of Burgundy. Of a similar character are 'lightning trips abroad' to foreign capitals which (p. 17) 'are not intended to provide the tourist with any profound "insights" but rather to give him a good time and have him return home "starry-eyed"'. The WTO also foresees an expansion of short-distance, short-stay rest and relaxation centres, which will variously provide entertainment or active sporting opportunities. Frechtling (1987), however, draws attention to other activities which may compete with pleasure travel for discretionary time and dollars. He notes, in particular, a rise in the role of the American home as an entertainment centre through such innovations as satellite dish television reception, video recording and playing and home computers.

New tourist demands are likely to complement rather than replace many of the well established existing patterns, such as mass tourism based on short- and medium-haul travel to coastal destinations, although changes may also be seen there such as a continuation of the move to more self-catering accommodation. In the future, the overall market for tourism will not only be larger, it will also be more complex and competitive. For these reasons increasing attention must be paid to market-segmentation and to matching supply and demand.

Shrinkages in more traditional agricultural and industrial sectors, burgeoning balance-of-payments deficits and lengthening unemployment lines are all factors which will ensure developers' interests in responding to these new demands. While illusory economic benefits continue to be seen by some, for example by politicians overestimating the job creation potential of tourism, there is, however, now often a more realistic appraisal of the economic impact of tourism and the contribution it might make to development. There has also been a growing awareness amongst the public sector, developers and host populations of associated social/cultural and environmental costs and benefits.

Consequently, the conditions in which tourism will develop in the future will differ from those experienced in the 1960s and 1970s. Fewer opportunities will exist for unrestrained, uncontrolled tourist development, with relatively unimpeded access to the best sites no longer going hand-in-hand with generous government subsidies or tax breaks. Developers, particularly external developers, will be faced by more stringent conditions, calls for more local participation and occasionally by well organized and informed opposition to their projects. To meet this changing situation and compete effectively in new markets, tourist development in the future must be better planned, more professionally managed and set in a broader context of development. Earlier chapters, particularly Ch. 7, have indicated such improvements are possible, though many obstacles still lie in the way. If, through stimulating greater awareness of these issues, providing examples of good and less desirable forms of tourist development and outlining appropriate concepts, methods and techniques, this book can contribute to further improvements in this field, then its aims will have been achieved.

References

ACAU (1967) *Contribution à l'Étude des Programmes de Nouvelles Stations de Vacances,* Les Cahiers du Tourisme, C–7, CHET, Aix-en-Provence.

Acerenza, M. A. (1985) Planificación estratégica del turismo: esquema metodológico, *Estudios Turisticos,* **85,** 47–70.

Ahmed, S. A. (1984) *Perceptions of Socio-economic and Cultural Impact of Tourism in Sri Lanka – A Research Study,* Working Paper 84–18, Faculty of Administration, University of Ottawa, Ottawa.

Airey, D. (1983) European government approaches to tourism, *Tourism Management,* **4**(4), 234–44.

Allcock, J. B. (1986) Yugoslavia's tourist trade: pot of gold or pig in a poke?, *Annals of Tourism Research,* **13**(4), 565–88.

An Foras Forbatha (1970) *Planning for Amenity, Recreation and Tourism,* An Foras Forbatha, Dublin.

An Foras Forbatha (1973) *Brittas Bay: a Planning and Conservation Study,* An Foras Forbatha, Dublin.

Andric, N. *et al.* (1962) Aspects régionaux de la planification touristique, *Tourist Review,* **17**(4), 230–6.

Andronicou, A. (1983) Selecting and planning for tourists – the case of Cyprus, *Tourism Management,* **4**(3), 209–11.

Archer, B. (1973) *The Impact of Domestic Tourism,* Bangor Occasional Papers in Economics No. 2, University of Wales Press, Bangor.

Archer, B. (1977) *Tourism Multipliers: The State of the Art,* Bangor Occasional Papers in Economics No. 11, University of Wales Press, Bangor.

293

Archer, B. H. (1980) Forecasting demand: quantitative and intuitive techniques, *Tourism Management,* 1(1), 5–12.

Archer, B. H. (1982) The value of multipliers and their policy implications, *Tourism Management,* 3(4), 236–41.

Archer, B. and **Shea, S.** (1973) *Gravity Models and Tourist Research,* Tourist Research Paper TUR2, Economics Research Unit, Bangor.

Archer, E. (1984) Estimating the relationship between tourism and economic growth in Barbados, *J. Travel Research,* 22(4), 8–12.

Archer, E. (1985) Emerging environmental problems in a tourist zone: the case of Barbados, *Caribbean Geography,* 2(1), 45–55.

Ascher, B. (1984) Obstacles to international travel and tourism. *J. Travel Research,* 22(3), 2–16.

Ashworth, G. J. and **Bergsma, J. R.** (1987) New policies for tourism: opportunities or problems?, *Tijdschrift voor Economische en Sociale Geografie,* 78(2), 151–5.

Ashworth, G. J. and **de Haan, T. Z.** (1986) *Uses and Users of the Tourist–Historic City: an Evolutionary Model in Norwich,* Serie Veldstudies No. 10, Geografisch Instdituut Rijksuniversiteit Groningen, Groningen.

Attanayake, A., Samaranayake, H. M. S. and **Ratnapala, N.** (1983) Sri Lanka, pp. 241–351 in Pye, E. A. and Lin, T.-B. (eds.), *Tourism in Asia: The Economic Impact,* Singapore University Press.

Baiderin, V. V. (1978) Effect of winter recreation on the soil and vegetation of slopes in the vicinity of Kazan, *Soviet J. Ecology,* 9(1), 76–86.

Baines, G. B. K. (1987) Manipulation of islands and men: sand–cay tourism in the South Pacific, pp. 16–24 in Britton, S. and Clarke, W. C. (eds.), *Ambiguous Alternative: Tourism in Small Developing Countries,* University of the South Pacific, Suva.

Barbaza, Y. (1970) Trois types d'intervention du tourisme dans l'organisation de l'espace littoral, *Annales de Géographie,* 434, 446–69.

Barbier, B. (1977) Les résidences secondaires et l'espace rural français, *Norois,* 96, 5–8.

Barbier, B. (1978) Ski et stations de sports d'hiver dans le monde, *Weiner Geographische Schriften,* 51/52, 130–46.

Barbier, B. and **Pearce, D. G.** (1984) The geography of tourism in France: definition, scope and themes, *GeoJournal,* 9(1), 47–53.

Baretje, R. (1982) Tourism's external account and the balance of payments, *Annals of Tourism Research,* 9(1), 57–67.

Baretje, R. and **Defert, P.** (1972) *Aspects Économiques du Tourisme,* Berger, Levrault, Paris.

Barker, M. L. (1982) Traditional landscape and mass tourism in the Alps, *Geographical Review,* 72(4), 395–415.

Barkham, J. P. (1973) Recreational carrying capacity: a problem of perception, *Area*, **5**, 218–22.

Barrett, J. A. (1958) *The Seaside Resort Towns of England and Wales*, Unpublished Ph.D. Thesis, University of London.

Bastin, R. (1984) Small island tourism: development or dependency?, *Development Policy Review*, **2**, 79–90.

Baud-Bovy, M. (1985) Bilan et Avenir de la Planification Touristique, *Cahiers du Tourisme C71*, Centre des Hautes Études Touristiques, Aix-en-Provence.

Bayfield, N. G. (1974) Burial of vegetation by erosion material near chairlifts on Cairngorm, *Biological Conservation*, **6**, 246–51.

Bayfield, N. G. (1979) Recovery of four montane communities on Cairngorm, Scotland, from disturbance by trampling, *Biological Conservation*, **16**(3), 165–79.

Belisle, F. J. and **Hoy, D. R.** (1980) The perceived impact of tourism by residents: a case study in Santa Marta, Colombia, *Annals of Tourism Research*, **7**(1), 83–101.

Bell, M. (1977) The spatial distribution of second homes: a modified gravity model, *J. Leisure Research*, **9**(3), 225–32.

Bello, D. C. and **Etzel, M. J.** (1985) The role of novelty in the pleasure travel experience, *J. Travel Research*, **24**(1), 20–6.

Ben Salem, L. (1970) Aspects humains du développement du tourisme dans le Cap Bon, *Revue Tunisienne des Sciences Sociales*, **20**.

Benthien, B. (1984) Recreational geography in the German Democratic Republic, *GeoJournal*, **9**(1), 59–63.

BERL (1982) *Tourism in the New Zealand Economy: A Summary*, New Zealand Tourist and Publicity Department, Wellington.

Berriane, M. (1978) Un type d'espace touristique marocain: le littoral méditerranéen, *Revue de Géographie du Maroc*, **29**(2), 5–28.

Berriane, M. (1986) Le tourisme et la petite ville au Maroc, *URBAMA*, *b16–17*, 187–207.

Besancenot, J.-P. (1985) Climat et tourisme estival sur les côtes de la péninsule ibérique, *Revue Géographique des Pyrénées et du Sud-Ouest*, **56**(4), 427–51.

Besancenot, J. P., Mounier, J. and **de Lavenne, F.** (1978) Les conditions climatiques du tourisme littoral: une méthode de recherche comprehensive, *Norois*, **99**, 357–82.

Bisson, J. (1986) *À l'Origine du Tourisme aux îles Baleares: Vocation Touristique ou Receptivité du Milieu d'Accueil?*, Paper presented at the meeting of the IGU Commission of the Geography of Tourism and Leisure, Palma de Mallorca (mimeo).

Blanchet, G. (1981) *Les Petites et Moyennes Entreprises Polynesiennes: le Cas de la petite Hôtellerie*, Travaux et Documents de l'ORSTOM, **136**, ORSTOM, Paris.

Boaglio, M. (1973) *Tourisme et Développement Economique en Espagne*, La

Documentation Française, Notes et Études Documentaires, 4048, Paris.

Boerjan, P. (1984) Les vacances des Belges en 1982, *Cahiers du Tourisme B31*, Centre des Hautes Études Touristiques, Aix-en-Provence.

Boerjan, P. and **Vanhove, N.** (1984) The tourism demand reconsidered in the context of the economic crisis, *Tourist Review*, 38(2), 2–11.

Boissevain, J. (1977) Tourism and development in Malta, *Development and Change*, 8, 523–38.

Booms, B. H. and **Bitner, M. J.** (1980) New management tools for the successful tourism manager, *Annals of Tourism Research*, 7(3), 337–52.

Boon, M. A. (1984) Understanding skier behaviour: an application of benefit segmentation market analysis to commercial recreation, *Loisir et Société*, 7(2), 397–406.

Boschken, H. L. (1975) The second home subdivision, market, suitability for recreational and pastoral use, *J. Leisure Research*, 7(1), 63–72.

Bounds, J. H. (1978) The Bahamas tourism industry: past, present and future, *Revista Geográfica*, 88, 167–219.

Bouquet, M. and **Winter, M.** (1987) *Who From Their Labours Rest? Conflict and Practice in Rural Tourism*, Avebury, Aldershot.

Bowles, J. M. and **Maun, M. A.** (1982) A study of the effects of trampling on the vegetation of Lake Huron – sand dunes at Pinery Provincial Park, *Biological Conservation*, 24(4), 273–83.

Boyer, J.-C. (1980) Résidences secondaires et 'rurbanisation' en région parisienne, *Tijdschrift voor Economische en Sociale Geografie*, 71(2), 78–87.

Braddon, C. J. H. (1982) *British Issues Paper: Approaches to Tourism Planning Abroad*, British Tourist Authority, London.

Britton, R. A. (1977) Making tourism more supportive of small state development: the case of St Vincent, *Annals of Tourism Research*, 6(5), 268–78.

Britton, S. G. (1980a) A conceptual model of tourism in a peripheral economy, pp. 1–12 in Pearce, D. G. (ed.), *Tourism in the South Pacific: the Contribution of Research to Development and Planning*, NZMAB Report No. 6, NZ National Commission for Unesco/Department of Geography, University of Canterbury, Christchurch.

Britton, S. G. (1980b) The spatial organisation of tourism in a neo-colonial economy: a Fiji case study, *Pacific Viewpoint*, 21(2), 144–65.

Britton, S. G. (1982) The political economy of tourism in the Third World, *Annals of Tourism Research*, 9(3), 331–58.

Britton, S. (1987) Tourism in Pacific-Island States: constraints and opportunities, pp. 113–39, in Britton, S. and Clarke, W. C. (eds.), *Ambiguous Alternative: Tourism in Small Developing Countries*, University of the South Pacific, Suva.

Bromberger, C. and Ravis-Giordani, G. (1977) *La Deuxième Phylloxera? Facteurs, Modalités et Conséquences de Migrations de Loisirs dans la Région Provence-Côte d'Azur. Étude Comparée de Quelques Cas*, Service Régional de l'Equipement/CETE, Aix-en-Provence.

Brown, I. W. *et al.* (1985), Monitoring sand dune erosion on the Clwyd Coast, North Wales, *Landscape Research*, **10**(3), 14–17.

Bryden, J. (1973) *Tourism and Development: a Case Study of the Commonwealth Caribbean*, Cambridge University Press, New York.

BTA (1976) *International Tourism and Strategic Planning*, British Tourist Association, London.

BTA (1982) *Employment in Tourism*, British Tourist Authority, London.

Buck, R. C. (1977) The ubiquitous tourist brochure: explorations in its intended and unintended use, *Annals of Tourism Research*, **4**(4), 195–207.

Burfitt, A. (1983) Research in Australian Tourism Commission Marketing, pp. 65–72 in *1983 PATA Travel Research Conference Proceedings*, Pacific Area Travel Association, San Francisco.

Burkart, A. J. (1971) Package holidays by air, *Tourist Review 1971*, **26**(2), 54–64.

Burkart, A. J. and Medlik, S. (1974) *Tourism: Past, Present and Future*, Heinemann, London.

Burnet, L. (1963) *Villégiature et Tourisme sur les Côtes de France*, Hachette, Paris.

Burtenshaw, D., Bateman, M. and Ashworth, G. J. (1981) *The City in West Europe*, Wiley, Chichester.

Butler, R. W. (1980) The concept of a tourist area cycle of evolution; implications for management of resources, *Canadian Geographer*, **24**(1), 5–12.

Cadart, C. (1975) *Les Nouvelles Implantations Hôtelières à Paris et dans la Région Parisienne*, Mémoire de Maitrise (unpublished), Centre d'Études Supérieures de Tourisme, Paris.

Calantone, R. J. and Johar, J. S. (1984) Seasonal segmentation of the tourism market using a benefit segmentation framework, *J. Travel Research*, **23**(2), 14–24.

Cals, J. (1974) *Turismo y Politica Turistica en España: una aproximación*, Editorial Ariel, Barcelona.

Cals, J., Esteban, J. and Teixidor, C. (1977) Les processus d'urbanisation touristique sur la Costa Brava, *Revue Géographique des Pyrénées et du Sud-Ouest*, **48**, 199–208.

Candela, R.-M. (1982) Piste de ski et érosion anthropique dans les Alpes du Sud, *Méditerranée*, (3–4), 51–5.

Carlson, A. S. (1938) Recreation industry of New Hampshire, *Economic Geography*, **14**, 255–70.

Carvajal, B. and **Patri, J.** (1979) Principios basicos para la obtención de un indice de jerarquización turistica, aplicado a la provincia Antarctica Chilena, *Inform. Geogr. Chile*, **26**, 65–80.

Casti Moreschi, E. (1986) *L'Espace Touristique Balnéaire Vénitien: Facteurs de Localisation et Dynamique Évolutive*, Paper presented at the meeting of the IGU Commission of the Geography of Tourism and Leisure, Palma de Mallorca (mimeo).

Cavallaro, C. (1985) Il turismo nella politica di piano in Thailandia, *Rassegna di Studi Turistici*, **20**(3–4), 247–71.

Cazes, G. (1972) Reflexions sur l'aménagement touristique du littoral du Languedoc-Roussillon, *L'Espace Géographique*, **1**(3), 193–210.

Cazes, G. (1978) Planification touristique et aménagement du territoire – les grandes tendances pour les années 80, pp. 76–88 in *Tourism Planning for the Eighties*, Editions AIEST, Berne.

Cazes, G. (1980a) Les avances pionnières du tourisme international dans le Tiers-Monde: réflexions sur un système décisionnel multinational en cours de constitution, *Travaux de l'Institut de Géographie de Reims*, **43–4**, 15–26.

Cazes, G. (1980b) *Les Aménagements Touristiques au Mexique*, Études et Mémoires 38, Centre des Hautes Etudes Touristiques, Aix-en-Provence.

Cazes, G. (1986) *Le Tourisme Alternatif: Réflexion sur un Concept Ambigu*, paper presented at Colloque du CNRS sur le Tourisme, Paris, 15–18 June 1986 (mimeo).

CECOD (1983) *Le Tourisme en France: Edition 1983 Réactualisé*, Centre d'Étude du Commerce et de la Distribution, Paris.

Centre on Transnational Corporations (1982) *Transnational Corporations in International Tourism*, United Nations, New York.

Chamberlain, R. N. (1983) Scheveningen, The Hague: the revitalisation of a declining holiday resort, pp. 25–33 in *Developing Tourism*, PRTC Education and Research Services, London.

Chappis, L. (1974) La montagne, où en est-on?, *Urbanisme*, **145**, 54–5.

Chenery, R. (1979) *A Comparative Study of Planning Considerations and Constraints Affecting Tourism Projects in the Principal European Capitals*, British Travel Educational Trust, London.

Chew, J. (1987) Transport and tourism in the year 2000, *Tourism Management*, **8**(2), 83–5.

Chmura, G. L. and **Ross, N. W.** (1978) *The Environmental Impacts of Marinas and their Boats: a Literature Review with Management Considerations*, Marine Memorandum 45, University of Rhode Island, Narragansett.

Choy, D. J. L. (1984) Forecasting tourism revisited, *Tourism Management*, **5**(3), 171–6.

Ciaccio, C. (1975) Développement touristique et groupes de pression en Sicile, *Travaux de l'Institut de Géographie de Reims*, **23–4**, 81–7.

Clary, D. (1984) Tourisme, urbanisation et organisation de l'espace: le littoral Bas-Normand, *Bulletin de l'Association des Géographes Français*, **501**, 125–31.

Clement, H. G. (1961) *The Future of Tourism in the Pacific and Far East*, Checchi and Coy, Washington DC.

Cleverdon, R. and Edwards, E. (1982) *International Tourism to 1990*, Abt Books, Cambridge.

Clout, H. D. (1972) Second homes in the U.S.A., *Tijdschrift voor Economische en Sociale Geografie*, **63**, 393–401.

Cohen, E. (1972) Toward a sociology of international tourism, *Social Research*, **39**, 164–82.

Cohen, E. (1974) Who is a tourist?, *Sociological Review*, **22**(4), 527–53.

Cohen, E. (1978) Impact of tourism on the physical environment, *Annals of Tourism Research* **5**(2), 215–37.

Cohen, E. (1982) Thai girls and farang men: the edge of ambiguity, *Annals of Tourism Research*, **9**(3), 403–28.

Cohen, E. (1984) The sociology of tourism: approaches, issues and findings, *Annual Review of Sociology*, **10**, 373–92.

Cole, D. N. (1987) Effects of three seasons of experimental trampling on five mountain forest communities and a grassland in Western Montana, U.S.A., *Biological Conservation*, **40**, 219–44.

Collins, C. O. (1979) Site and situation strategy in tourism planning: a Mexican case study, *Annals of Tourism Research*, **6**(3), 351–66.

Cooke, K. (1982) Guidelines for socially appropriate tourism development in British Columbia, *J. Travel Research*, **21**(1), 22–8.

Cooper, M. J. and Pigram, J. J. (1984) Tourism and the Australian economy, *Tourism Management*, **5**(1), 2–12.

Coppock, J. T. (ed.) (1977) *Second Homes: Curse or Blessing?*, Pergamon, London.

Courbis, R. (1984) Les conséquences économiques et sociales de l'aménagement touristique: l'impact des investissements réalisés dans le cas du Languedoc–Roussillon, *World Travel*, **180**, 29 and 33.

Cribier, F. (1973) Les résidences secondaires des citadins dans les campagnes françaises, *Études Rurales*, **49–50**, 181–204.

Crompton, J. L. (1979) Motivations for pleasure vacation, *Annals of Tourism Research*, **6**(4), 408–24.

Crowe, R. B. (1975) Recreation, tourism and climate – a Canadian perspective, *Weather*, **30**(8), 248–54.

Crozier, M. J., Marx, S. L. and Grant, I. J. (1978) Off-road vehicle recreation: the impact of off-road motorcycles on soil and vegetation conditions, pp. 76–9, *Proc. 9th N.Z. Geog. Conf.*, N.Z. Geog. Soc., Dunedin.

Crush, J. S. and **Wellings, P. A.** (1987) Forbidden fruit and the export of vice: tourism in Lesotho and Swaziland, pp. 91–112 in Britton, S. and Clarke, W. C. (eds.), *Ambiguous Alternative: Tourism in Small Developing Countries*, University of the South Pacific, Suva.

Cumin, G. (1966) Capacité du domaine skiable, *Économie et Prospective de la Montagne*, **7**, 20–4.

Cumin, G. (1970) Les stations intégrées, *Urbanisme*, **116**, 50–3.

Dahl, A. L. (1981) *Coral Reef Monitoring Handbook*, South Pacific Commission, Noumea.

Dale, D. and **Weaver, T.** (1974) Trampling effects on vegetation of the trail corridors of north Rocky Mountain forests, *J. Applied Ecology*, **11**, 767–72.

D'Amore, L. J. (1983) Guidelines to planning in harmony with the host community, pp. 135–59 in Murphy, P. E. (ed.), *Tourism in Canada: Selected Issues and Options*, Western Geographical Series, Vol. 21, University of Victoria, Victoria.

Dann, G. M. S. (1977) Anomie, ego-enhancement and tourism, *Annals of Tourism Research*, **4**(4), 184–94.

Dartington Amenity Research Trust (1974) *Farm Recreation and Tourism in England and Wales; A Report to the Countryside Commission, English Tourist Board and Wales Tourist Board*, Dart Publication No. 14.

Dauphiné, A. and **Ghilardi, N.** (1978) Essai de bioclimatologie touristique: La Côte d'Azur, *Méditerranée*, **33**(3), 3–15.

Davies, E. (1987) Planning in the New Zealand national park, *New Zealand Geographer*, **43**(2), 73–8.

Davis, B. D. and **Sternquist, B.** (1987) Appealing to the elusive tourist: an attribute cluster strategy, *J. Travel Research*, **25**(4), 25–31.

Davis, H. D. and **Simmons, J. A.** (1982) World Bank experience with tourism projects, *Tourism Management*, **3**(4), 212–17.

Davis, H. L. and **Rigaux, B. P.** (1974) Perception of marital roles in decision processes, *J. Consumer Research*, **1**(1), 51–62.

Day, E. E. D., McCalla, R. J., Millward, H. A. and **Robinson, B. S.** (1977) *The Climate of Fundy National Park and its Implications for Recreation and Park Management*, Atlantic Region Geographical Series No. 1, Department of Geography, Saint Mary's University, Halifax.

Dean, J. and **Judd, B.** (eds) (1985) *Tourist Developments in Australia*, RAIA Education Division, Red Hill.

Deasy, G. F. (1949) The tourist industry in a North Woods county, *Economic Geography*, **25**(2), 240–59.

Defert, P. (1966) Le tourisme: facteur de valorisation régional, *Recherche Sociale*, **3**.

Defert, P. (1972) *Les Ressources et les Activités Touristiques: Essai* *d'Intégration*, Les Cahiers du Tourisme, C–19, CHET, Aix-en-Provence.

Dernoi, L. A. (1983) Farm tourism in Europe, *Tourism Management*, 4(3), 155–66.

Din, K. H. (1982) Tourism in Malaysia: competing needs in a plural society, *Annals of Tourism Research*, 9(3), 453–80.

van Doorn, J. W. M. (1979) *The Developing Countries: Are they Really Affected by Tourism? Some Critical Notes on Socio-cultural Impact Studies*, Paper presented at Seminar on Leisure Studies and Tourism, 7–8 December 1974, Warsaw (mimeo).

van Doorn, J. W. M. (1986) Scenario writing: a method for long-term forecasting, *Tourism Management*, 7(1), 33–49.

Dorfmann, M. (1983) Régions de montagne: de la dépendance à l'autodéveloppement, *Revue de Géographie Alpine*, 71 (1), 5–34.

Doxey, G. V. (1975) A causation theory of visitor–resident irritants: methodology and research inferences, pp. 195–8 in *The Impact of Tourism, Sixth Annual Conference Proceedings*, Travel Research Association, San Diego.

Duffield, B. S. (1982) Tourism: the measurement of economic and social impact, *Tourism Management*, 3(4), 248–55.

Duffield, B. S. (1984) The study of tourism in Britain – a geographical perspective, *GeoJournal*, 9(1), 27–35.

Duffield, B. S. and **Long, J.** (1981) Tourism in the Highlands and Islands of Scotland: rewards and conflicts, *Annals of Tourism Research*, 8(3), 403–31.

Duffield, B. S. and **Long, J.** (1984) The role of tourism in the economy of Scotland, *Tourism Management*, 5(4), 258–68.

Dumas, D. (1975a) Un type d'urbanisation touristique littorale: la Manga del Mar Menor (Espagne), *Travaux de l'Institut de Géographie de Reims*, 23–4, 89–96.

Dumas, D. (1975b) Evolution demographique récente et développement du tourisme dans la province d'Alicante, Espagne, *Méditerranée*, 21, 3–22.

Dumas, D. (1976) L'urbanisation touristique du littoral de la Costa Blanca (Espagne), *Cahiers Nantais*, 13, 43–50.

ECAC (1981) *Statistics on Non-Scheduled Traffic Reported in ECAC States, 1980*, European Civil Aviation Conference, Paris.

Edington, J. M. and **Edington, M. A.** (1977) *Ecology and Environmental Planning*, Chapman Hall, London.

Edington, J. M. and **Edington, M. A.** (1986) *Ecology, Recreation and Tourism*, Cambridge University Press, Cambridge.

Edwards, A. (1985) International tourism forecasts to 1995, *International Tourism Quarterly*, (2), 52–64.

Elkan, W. (1975) The relation between tourism and employment in

Kenya and Tanzania, *J. Development Studies*, **11**(2), 123–30.

England, J. L., Gibbons, W. E. and **Johnson, B. L.** (1980) The impact of ski resorts on subjective well-being, *Leisure Sciences*, **3**(4), 311–48.

Entwistle, E. R. (1987) Methods of economic evaluation of national parks with reference to New Zealand, *New Zealand Geographer*, **43**(2), 79–83.

Erisman, M. (1983) Tourism and cultural dependency in the West Indies, *Annals of Tourism Research*, **10**(3), 337–62.

Ersek, S. and **Düzgunoglu, E.** (1976) An approach to physical planning for tourism development in Turkey, pp. 68–76 in ECE, *Planning and Development of the Tourist Industry in the ECE Region*, United Nations, New York.

Esteban Talaya, A. and **Figuerola Paloma, M.** (1984) Técnicas de previsión y análisis de comportamiento de la demanda turistica, *Estudios Turisticos*, **84**, 3–16.

ETB (1981) *Tourism and the Inner City*, English Tourist Board, London.

Etzel, M. J. and **Woodside, A. G.** (1982) Segmenting vacation markets: the case of distant and near-home travellers, *J. Travel Research*, **20**(4), 1982, 10–14.

European Omnibus Survey (1986) *Europeans' Holidays: a summary of the results of a survey conducted in March/April 1986 in the twelve countries of the European Community*, Commission of the European Communities, Brussels.

Ewing, G. O. and **Kulka, T.** (1979) Revealed and stated preference analysis of ski resort attractiveness, *Leisure Sciences*, **2**, 249–75.

FAO/ECE (1982) *Report on the Symposium on Agriculture and Tourism Mariehamn, 7th to 12th of June 1982*, Government Printing Centre, Helsinki.

Farrell, B. H. (1982) *Hawaii, the Legend that Sells*, University Press of Hawaii, Honolulu.

Farwell, T. A. (1970) Resort planning and development, *Cornell H. R. A. Quarterly*, February, 34–7.

Federal Department of Forestry (1981) A case study of the growth of ski tourism and environmental stress in Switzerland, pp. 261–318 in *Case Studies on the Impact of Tourism on the Environment*, OECD, Paris.

Fernandez Fuster, L. (1974) *Teoria y Técnica del Turismo*, 4th edn, Editora Nacional, Madrid.

Ferrario, F. F. (1979) The evaluation of tourist resources: an applied methodology, *J. Travel Research*, **17**(3), 18–22 and **17**(4), 24–30.

Fesenmaier, D. R. and **Roehl, W. S.** (1986) Locational analysis in campground development decisions, *J. Travel Research*, **14**(3), 18–22.

Figuerola, M. (1976) Turismo de masa y sociologia: el caso español, *Travel Research J.*, 25–38.

Fish, M. M. (1984) On controlling sex sales to tourists: commentary on Graburn and Cohen, *Annals of Tourism Research*, 11(4), 615–17.

Foin, T. C. *et al.* (1981) Quantitative studies of visitor impacts on environments of Yosemite National Park, California, and their implications for park management policy. *J. Environmental Management*, 5(1), 1–22.

Ford, L. R. (1979) Urban preservation and the geography of the city in the USA, *Progress in Human Geography*, 3(2), 211–38.

Forer, P. C. and **Pearce, D. G.** (1984) Spatial patterns of package tourism in New Zealand, *New Zealand Geographer*, 40(1), 34–42.

Forster, J. (1964) The sociological consequences of tourism, *International Journal of Comparative Sociology*, 5(2), 217–27.

Fortin, P. A. and **Ritchie, J. R. B.** (1977) An empirical study of association decision processes in convention site selection, *J. Travel Research*, 15(4), 13–20.

Francisco, R. (1983) The political impact of tourism dependence in Latin America, *Annals of Tourism Research*, 10(3), 363–76.

Frater, J. M. (1983) Farm tourism in England. Planning, funding, promotion and some lessons from Europe, *Tourism Management*, 4(3), 167–79.

Frechtling, D. C. (1987) Key issues in tourism futures. The US travel industry, *Tourism Management*, 8(2), 106–11.

Freeman, A. M. (1975) A survey of the techniques for measuring the benefits of water quality improvement, in Peskin, H. M. and Seskin, E. P. (eds.), *Cost–Benefit Analysis and Water Pollution Policy*, The Urban Institute, Washington.

Friedmann, J. (1980) An alternative development, and communalistic society: some principles for a possible future, pp. 12–42 in Friedmann, J, Wheelwright, E. and Connell, J. *Development Strategies in the Eighties*, Development Studies Colloquium, Monograph No. 1, Sydney.

Gade, D. W. (1986) *Tropicalization of Mediterranean Lands: Ornamental Vegetation on the French Riviera*, Paper presented at the International Geographical Union's Regional Conference on Mediterranean Lands, Barcelona (mimeo).

Garland, B. R. and **West, S. J.** (1985) The social impact of tourism in New Zealand, *Massey Journal of Asian and Pacific Business*, 1(1), 34–9.

Gaviria, M. (1975) *Turismo de Playa en España*, Ediciones Turner, Madrid.

Gaviria, M. *et al.* (1974) *España a Go-Go. Turismo Charter y Neocolonialismo del Espacio*, Ediciones Turner, Madrid.

Gearing, C. E. and **Var, T.** (1977) Site selection problem in touristic feasibility reports, *Tourist Review*, **32**(2), 9–16.

Gee, C. (1984) Fifth General Assembly of the World Tourism Organisation, *Annals of Tourism Research*, **11**(2), 305–6.

Georgulas, N. (1970) Tourist destination features, *J. Town Planning Institute*, **10**, 442–6.

Getz, D. (1986a) Tourism and population change: long-term impacts of tourism in the Badenoch and Strathspey district of the Scottish highlands, *Scottish Geographical Magazine*, **102**(2), 113–26.

Getz, D. (1986b) Models in tourism planning, *Tourism Management*, **7**(1), 21–32.

Getz, D. (1987) *Tourism Planning and Research Traditions, Models and Futures,* Paper presented at Australian Travel Research Workshop, Bunbury, 5–6 November 1987 (mimeo).

Ghazanshahi, J., Huchel, T. D. and **Devinny, J. S.** (1983) Alteration of Southern Californian rocky shore ecosystems by public recreational use, *J. Environmental Management*, **16**, 379–94.

Gilbert, E. W. (1939) The growth of inland and seaside health resorts in England, *Scottish Geographical Magazine*, **55**, 16–35.

Giourgas, G. (1985) Profil des voyageurs européens et leurs attitudes vis-à-vis l'unification de l'Europe, pp. 39–67 in Cerexhe, E. and Giourgas, G. (eds.), *Tourisme et Intégration Européene*, Ciaco, Brussels.

Giraud, G. (1971) Port Grimaud, a French water-town under construction, *Build International*, May/June, 144–7.

Gitelson, R. J. and **Crompton, J. L.** (1983) The planning horizons and sources of information used by pleasure vacationers, *J. Travel Research*, **21**(3), 2–7.

Goodrich, J. N. (1977) Differences in perceived similarity of tourism regions: a spatial analysis, *J. Travel Research*, **16**(1), 10–13.

Goonatilake, S. (1978) *Tourism in Sri Lanka: the Mapping of International Inequalities and their Internal Structural Effects,* Working Paper No. 19, Centre for Developing-Area Studies, McGill University, Montreal.

Gorman, M. *et al.* (eds.) (1977) *Design for Tourism, an ICSID Interdesign Report,* Pergamon, Oxford.

Gormsen, E. (1981) The spatio-temporal development of international tourism: attempt at a centre–periphery model, pp. 150–70 in *La Consommation d'Espace par le Tourisme et sa Preservation*, CHET, Aix-en-Provence.

Gottlieb, A. (1982) Americans' vacations, *Annals of Tourism Research*, **9**(2), 165–87.

Goulet, G. (1968) On the goals of development, *Cross Currents*, **18**, 387–405.

Graburn, N. H. H. (1983a) The anthropology of tourism, *Annals of Tourism Research*, **10**(1), 9–33.

Graburn, N. H. H. (1983b) Tourism and prostitution, *Annals of Tourism Research*, **10**(3), 437–43.

Graham, J. E. J. and Wall, G. (1978) American visitors to Canada: a study in market segmentation, *J. Travel Research*, **16**(3), 21–4.

Gray, F. and Lowerson, J. (1979) Seaside see-saw, *Geographical Magazine*, **51**(6), 433–38.

Gray, H. P. (1970) *International Travel – International Trade*, D. C. Heath and Company, Lexington.

Gray, H. P. (1982) The contributions of economics to tourism, *Annals of Tourism Research*, **9**(1), 105–25.

Greenland, D. E. (1979) *Modelling Air Pollution Potential for Mountain Resorts*, Occasional Paper No. 32, Institute of Arctic and Alpine Research, University of Colorado, Boulder.

Greenwood, D. J. (1972) Tourism as an agent of change: a Spanish Basque case, *Ethnology*, **11**, 80–91.

Greer, T. and Wall, G. (1979) Recreational hinterlands: a theoretical and empirical analysis, pp. 227–45 in Wall, G. (ed.), *Recreational Land Use in Southern Ontario*, Department of Geography Publication Series, No. 14, University of Waterloo.

Grinstein, A. (1955) Vacations: a psycho-analytic study, *Int. J. Psycho-Analysis*, **36**(3), 177–85.

Guérin, J.-P. (1984) *L'Aménagement de la Montagne en France: Politiques, Discours et Productions d'Espace dans les Alpes du Nord*, Ophrys.

Guitart, C. (1982) UK charter flight package holidays to the Mediterranean, 1970–78: a statistical analysis, *Tourism Management*, **3**(1), 16–39.

Gunadhi, H. and Boey, C. K. (1986) Demand elasticities of tourism in Singapore, *Tourism Management*, **7**(4), 239–53.

Gunn, C.A. (1979) *Tourism Planning*, Crane Rusak, New York.

Gunn, C. A. and Jafari, J. (1980) World Tourism Conference: an inter-governmental tourism landmark, *Annals of Tourism Research*, **7**(3), 478–87.

Gutiérrez Ronco, S. (1977) Localización actual de la hosteleria madrileña, *Boletin de la Real Sociedad Geografica 1976*, part 2, 347–57.

Gutiérrez Ronco, S. (1980) Evolución en la localización de la hosteleria madrilena, pp. 283–8 in *Jornadas de Estudios sobre la Provincia de Madrid*, Diputación Provincial de Madrid, Madrid.

Haahti, A. J. (1986) Finland's competitive position as a destination, *Annals of Tourism Research*, **13**(1), 11–35.

Haider, W. (1985) Small accommodation development in the Caribbean: an appraisal, pp. 63–8 in Pulsipher, L. (ed.), *Proceedings of* *the Conference of Latin Americanist Geographers*, Ball State University, Muncie.

Hall, D. R. (1984) Foreign tourism under socialism: the Albanian "Stalinist" model, *Annals of Tourism Research*, **11**(4), 539–55.

Hall, P. (1970) *Theory and Practice of Regional Planning*, Pemberton Books, London.

Hanna, M. (1976) *Tourism Multipliers in Britain: a Guide to Economic Multipliers in Theory and Practice*, English Tourist Board, London.

Harker, J. (1973) Estimating the climatic potential for winter recreation, *Trent Student Geographer*, **2**, 38–46.

Hartmann, R. (1986) Tourism, seasonality and social change, *Leisure Studies*, **5**(1), 25–33.

Haukeland, J. V. (1984) Sociocultural impacts of tourism in Scandinavia: studies of three host communities, *Tourism Management*, **5**(3), 207–14.

Haulot, A. (1981) Social tourism: current dimensions and future developments, *Tourism Management*, **2**(3), 207–12.

Hawes, D. K. (1977) Psychographics are meaningful . . . not merely interesting, *J. Travel Research*, **25**(4), 1–7.

Heape, R. (1983) Tour operating planning in Thomson holidays UK: the use of research, *Tourism Management*, **4**(4), 245–52.

Hector, C. (1978) Tourisme, dépendence et division internationale du travail dans les Caraibes insulaires, *Manpower and Unemployment Research*, **11**(2), 3–8.

Heeley, J. (1981) Planning for tourism in Britain, *Town Planning Review*, **52**, 61–79.

Heeley, J. (1986a) 1986: a significant year of change for tourism organisation in Scotland? *Fraser of Allander Quarterly Economic Commentary*, **11**(3), 66–9.

Heeley, J. (1986b) Big company involvement in Scottish tourism, *Fraser of Allander Quarterly Economic Commentary*, **12**(2), 75–9.

Heeley, J. (1986c) Tourism in England: a strategic planning response?, pp. 22–30 in *Strategy and Opportunities for Tourism Development*, Planning Exchange Occasional Paper No. 22.

Henderson, D. M. (1975) *The Economic Impact of Tourism: A Case Study of Greater Tayside*, University of Edinburgh, TRRU, Research Report No. 13, Edinburgh.

Hermans, D. (1981) The encounter of agriculture and tourism: a Catalan case, *Annals of Tourism Research*, **8**(3), 462–79.

Hills, T. L. and **Lundgren, J.** (1977) The impact of tourism in the Caribbean: a methodological study, *Annals of Tourism Research*, **4**(5), 248–67.

Hodgson, P. (1983) Research into the complex nature of the holiday choice process, pp. 17–35 in *Proceedings of the Seminar on* *the Importance of Research in the Tourist Industry*, Helsinki, 8–11 June 1983, ESOMAR, Amsterdam.

Hoivik, T. and **Heiberg, T.** (1980) Centre–periphery tourism and self-reliance, *Int. Soc. Sci. J.* **32**(1), 69–98.

Hollander, G. (1982) *Determinants of Demand for Travel to and from Australia*, Working Paper No. 26, Bureau of Industry Economics, Canberra.

Holloway, J. C. (1985) *The Business of Tourism*, 2nd edn, MacDonald and Evans, Plymouth.

House, J. W. (1978) *France: An Applied Geography*, Methuen, London.

van Houts, D. (1983) Female participation in tourism employment in Tunisia: some economic and non-economic costs and benefits. *Tourist Review*, **38**(1), 25–30.

Hudman, L. E. (1980) *Tourism: a Shrinking World*, Grid Inc., Columbus.

Hughes, C. G. (1982) The employment and economic effects of tourism reappraised, *Tourism Management*, **3**(3), 167–76.

Hughes, H. L. (1984) Government support for tourism in the UK, *Tourism Management*, **5**(1), 13–19.

Hunt, J. D. (1975) Image as a factor in tourism development, *J. Travel Research*, **13**(3), 1–7.

Husbands, W. C. (1986) Periphery resort tourism and tourist-resident stress: an example from Barbados, *Leisure Studies*, **5**, 175–88.

Inskeep, E. (1987) Environmental planning for tourism, *Annals of Tourism Research*, **14**(11), 118–35.

Ireland, M. (1987) Planning policy and holiday homes in Cornwall, pp. 65–82 in Bouquet, M. and Winter, M. (eds.), *Who From Their Labours Rest?*, Avebury, Aldershot.

Iso-Ahola, S. E. (1982) Toward a social psychological theory of tourism motivation: a rejoinder, *Annals of Tourism Research*, **9**(2), 256–61.

ITQ (1984) UK travel agents – who they are and their market. Part 2: the agents, *International Tourism Quarterly*, 40–56.

IUOTO (1975) *The Impact of International Tourism on the Economic Development of the Developing Countries*, International Union of Official Travel Organizations/World Tourism Organization, Geneva.

Jackson, I. (1984) *Enhancing the Positive Impact of Tourism on the Built and Natural Environment*, Vol. 5, *Reference Guidelines for Enhancing the Positive Socio-cultural and Environmental Impacts of Tourism*, Organization of American States, Washington.

Jafari, J. (1982) Understanding the structure of tourism – an *avant propos* to studying its costs and benefits, pp. 51–67 in *Interrelations Between Benefits and Costs of Tourist Resources*, Association Internationale des Experts Scientifiques du Tourisme, Berne.

Jansen-Verbeke, M. (1986) Inner-city tourism: resources, tourists and promoters, *Annals of Tourism Research*, 13(1), 79–100.

Jansen-Verbeke, M. (1988) *Leisure, Recreation and Tourism in Inner Cities: Explorative Case-Studies*, Katholieke Universiteit Nijmegen, Nijmegen.

Jenkins, C. L. (1980) Tourism policies in developing countries: a critique, *Tourism Management*, 1(1), 22–9.

Jenkins, C. L. (1982) The effects of scale in tourism projects in developing countries, *Annals of Tourism Research*, 9(2), 229–49.

Jenkins, C. L. and Henry, B. M. (1982) Government involvement in tourism in developing countries, *Annals of Tourism Research*, 9(4), 499–521.

Jenkins, R. L. (1978) Family vacation decision-making, *J. Travel Research*, 26(4), 2–7.

Jocard, L.-M. (1965) *Le Tourisme et l'Action de l'État*, Berger-Levrault, Paris.

Joliffe, I. P. and Patman, G. R. (1986) Roses Bay, North East Spain: the impact of tourism on a Mediterranean coastal landscape, pp. 113–28 in *Contemporary Ecological-Geographical Problems of the Mediterranean*, International Geographical Union/UNESCO, Palma de Mallorca.

Johnstone, I. M., Coffey, B. T. and Howard-Williams, C. (1985) The role of recreational boat traffic in interlake dispersal of macrophytes: a New Zealand case study, *J. Environmental Management*, 20(3), 263–79.

Jones, D. R. W. (1978) *Prostitution and Tourism*, Paper presented to the Peacesat Conference on the Impact of Tourism Development in the Pacific (mimeo).

de Kadt, E, (1979) *Tourism: Passport to Development?*, Oxford University Press, Oxford.

Kassé, M. (1973) La théorie du développement de l'industrie, touristique dans les pays sous-développés, *Annales Africaines*, 1971–72, 53–72.

Kay, J. *et al.* (1981) Evaluating environmental impacts of off-road vehicles, *J. Geography*, 80(1), 10–18.

Kaynak, E. and Macaulay, J. A. (1984) The Delphi technique in the measurement of tourism market potential, *Tourism Management*, 5(2), 87–101.

Kaynak, E., Odabasi, Y. and Kavas, A. (1986) Tourism marketing in a developing economy: frequent and infrequent visitors contrasted, *Service Industries Journal*, 6(1), 53–60.

Keller, C. P. (1987) Stages of peripheral tourism development – Canada's Northwest Territories, *Tourism Management,* **8**(1), 20–32.

Keller, P. (1976) Objectives and measures for achieving planned tourist development, pp. 190–8, in ECE, *Planning and Development of the Tourist Industry in the ECE Region,* United Nations, New York.

Keogh, B. (1980) Motivations and the choice decisions of skiers, *Tourist Review,* **35**(1), 18–22.

Keogh, B. (1982) L'impact social du tourisme: le cas de Shédiac, Nouveau–Brunswick, *Canadian Geographer,* **26**(4), 318–31.

Kirkpatrick, L. W. and **Reeser, W. K.** (1976) The air pollution carrying capacities of selected Colorado mountain valley ski communities, *J. Air Pollution Control Association,* **26**(10), 992–4.

Kissling, C. C. (1980) *International Civil Aviation in the South Pacific: A Perspective,* Occasional Paper No. 19, Development Studies Centre, Australian National University, Canberra.

Klöpper, R. (1976) Physical planning and tourism in the Federal Republic of Germany, pp. 50–6 in ECE, *Planning and Development of the Tourist Industry in the ECE Region,* United Nations, New York.

Knafou, R. (1978) *Les Stations Intégrées de Sports d'Hiver des Alpes Françaises,* Masson, Paris.

Knafou, R. (1979) L'aménagement du territoire en économie libérale: l'exemple des stations intégrées de sports d'hiver des Alpes françaises, *L'Espace Géographique,* **8**(3), 173–80.

Kofman, E. (1985) Dependent development in Corsica, pp. 263–83 in Hudson, R. and Lewis, J. (eds.), *Uneven Development in Southern Europe,* Methuen, London and New York.

Krapf, K. (1961) Les pays en voie de développement face au tourisme. Introduction méthodologique, *Tourist Review,* **16**(3), 82–9.

Krippendorf, J. (1977) *Les Devoreurs des Paysages,* 24 Heures, Lausanne.

Lambiri-Dimaki, J. (1976) Tourism and cultural development: the undermining of a myth, pp. 282–5 in *Les Problèmes de Management dans le Domaine du Tourisme,* Editions Gurten, Berne.

Lanquar, R. and **Raynouard, Y.** (1978) *Le Tourisme Social,* Presses Universitaires de France, Paris.

Latimer, H. (1985) Developing-island economies – tourism v agriculture, *Tourism Management,* **6**(1), 32–42.

Lavery, P. (1974) The demand for recreation, pp. 22–48 in Lavery, P. (ed.), *Recreational Geography,* David and Charles, Newton Abbot.

Law, C. M. (1985) *Urban Tourism in the United States,* Working Paper No. 4, Department of Geography, University of Salford.

Lawson, F. and **Baud-Bovy, M.** (1977) *Tourism and Recreation Development,* Architectural Press, London.

Lee, G. P. (1987) Tourism as a factor in development cooperation, *Tourism Management,* 8(1), 2–19.

Lee, R. L. (1978) Who owns boardwalk? The structure of control in the tourist industry of the Yucatan, *Studies in Third World Societies,* 6, 19–35.

Lefevre, V. and **Renard, J.** (1980) Tourisme, agriculture et habitat dans l'intérieur des Marais de Monts, *Cahiers Nantais,* 18, 45–65.

Leiper, N. (1984) Tourism and leisure: the significance of tourism in the leisure spectrum, pp. 249–53, in *Proc. 12th N.Z. Geog. Conf.* N.Z. Geog. Soc., Christchurch.

Lever, A. (1987) Spanish tourism migrants: the case of Lloret de Mar, *Annals of Tourism Research,* 14(4), 449–70.

Lew, A. A. (1987) A framework of tourist attraction research, *Annals of Tourism Research,* 14(3), 553–75.

Lewis, R. C. (1985) The basis of hotel selection, *Cornell Hotel and Restaurant Administration Quarterly,* 25(1), 54–69.

Lichtenberger, E. R. (1984) Geography of tourism and leisure society in Austria, *GeoJournal,* 9(1), 41–6.

Liddle, M. J. (1975) A selective review of the ecological effects of human trampling on natural ecosystems, *Biological Conservation,* 7, 17–34.

Liddle, M. J. and **Scorgie, H. R. A.** (1980) The effects of recreation on freshwater plants and animals: a review, *Biological Conservation,* 17(3), 183–206.

Lindsay, J. J. (1986) Carrying capacity for tourism development in national parks of the United States, *Industry and Environment,* 9(1), 17–20.

Liu, J. C. (1983) Hotel industry performance and planning at the regional level, pp. 211–33 in Murphy, P. E. (ed.), *Tourism in Canada: Selected Issues and Options,* Western Geographical Series Vol. 21, University of Victoria, Victoria.

Liu, J. and **Var, T.** (1982) Differential multipliers for the accommodation sector, *Tourism Management,* 3(3), 177–87.

Liu, J. C. and **Var, T.** (1986) Resident attitudes toward tourism impacts in Hawaii, *Annals of Tourism Research,* 13(2), 193–214.

Liu, J., Var, T. and **Timur, A.** (1984) Tourist-income multipliers for Turkey, *Tourism Management,* 5(4), 280–7.

Loeb, P. D. (1982) International travel to the United States: an econometric evaluation, *Annals of Tourism Research,* 9(1), 7–20.

London, C. C. (1975) *Quality Skiing at Aspen, Colorado: A Study in Recreational Carrying Capacity,* Occasional Paper No. 14, Institute of Arctic and Alpine Research, Boulder.

Long, F. (1978) Tourism: a development cornerstone that crumbled: a case history from the Caribbean, *CERES,* 11(5), 43–5.

Lundberg, A. (1984) A controversy between recreation and ecosystem protection in the sand dune areas on Karmoy, Southwestern Norway, *GeoJournal*, **8**(2), 147–57.

Lundberg, D. E. (1974) *The Tourist Business*, 2nd edn, Cahners Books, Boston.

Lundgren, J. O. J. (1972) The development of tourist travel systems – a metropolitan economic hegemony par excellence, *Jahrbuch fur Fremdenverkehr*, **20** Jahrgang, 86–120.

Lundgren, J. O. J. (1974) On access to recreational lands in dynamic metropolitan hinterlands, *Tourist Review*, **29**(4), 124–31.

Lundgren, J. O. J. (1982) The tourist frontier of Nouveau Quebec: functions and regional linkages, *Tourist Review*, **37**(2), 10–16.

Lundgren, J. O. J. (1984) Geographic concepts and the development of tourism research in Canada, *GeoJournal*, **9**(1), 17–25.

Mabogunje, A. L. (1980) *The Development Process: A Spatial Perspective*, Hutchinson, London.

McAllister, D. M. and Klett, F. R. (1976) A modified gravity model of regional recreation activity with an application to ski trips, *J. Leisure Research*, **8**(1), 22–34.

McBoyle, G. and Wall, G. (1987) The impact of CO_2-induced warming on downhill skiing in the Laurentians, *Cahiers de Géographie du Québec*, **31**(82), 39–50.

McCann, B. (1983) The economic impact of tourism, pp. 7–18 in *Workshop on Measuring the Impacts of Tourism*, Pacific Area Travel Association, San Francisco.

McDermott, P. J. and Jackson, L. R. (1985) *The Economic Determinants of Tourist Arrivals in Australia and New Zealand*, New Zealand Tourist Industry Federation, Wellington.

McDonnell Douglas (1977) *The European Charter Airlines*, 2nd edn, McDonnell Douglas, Longbeach.

McEachern, J. and Towle, E. L. (1974) *Ecological Guidelines for Island Development*, International Union for the Conservation of Nature, Morges.

McElroy, J. L. and de Albuquerque, K. (1986) The tourism demonstration effect in the Caribbean, *J. Travel Research*, **25**(2), 31–4.

McIntosh, R. W. and Goeldner, C. R. (1986) *Tourism: Principles, Practices, Philosophies*, 5th edn, John Wiley, New York.

McTaggart, W. D. (1980) Tourism and tradition in Bali, *World Development*, **8**, 457–66.

McVey, M. and Heeley, J. (1984) Tourism and public policy in Scotland, *Fraser of Allander Quarterly Economic Commentary*, **10**(1), 63–70.

Malamud, B. (1973) Gravity model calibration of tourist travel to Las Vegas, *J. Leisure Research*, **5**(1), 23–33.

Marris, T. (1986) Does food matter, *Tourist Review*, **39**(4), 17–20.
Martinelli, M. (1976) Meteorology and ski area development and operation, in *Proceedings Fourth National Conference on Fire and Forest Meteorology*, USDA For. Serv. Gen. Tech. Rep. RM–32.
Mathieson, A. and **Wall, G.** (1982) *Tourism: Economic, Physical and Social Impacts*, Longman, London.
Maurer, J. L. (1980) *Tourism and Development in a Socio-cultural Perspective, Indonesia as a Case Study*, 2nd edn, Institut Universitaire d'Études du Développement, Geneva.
Mayo, E. J. (1974) A model of motel-choice, *Cornell Hotel and Restaurant Administration Quarterly*, **15**(3), 55–64.
Mazanec, J. A. (1984) How to detect travel market segments: a clustering approach, *J. Travel Research*, **23**(1), 17–21.
Meunier, G. (1985) Étude sur les motivations, les attentes et les habitudes des familles à faible revenu en regard des activités de loisir touristique, *Loisir et Société*, **8**(2), 513–25.
Meyer-Arendt, K. J. (1985) The Grand Isle, Louisiana resort cycle, *Annals of Tourism Research*, **12**(3), 449–65.
Michaud, J.-L. (1983) *Le Tourisme Face à l'Environnement*, Presses Universitaires de France, Paris.
Mieczkowski, Z. (1985) The tourism climatic index: a method of evaluating world climates for tourism, *Canadian Geographer*, **29**(3), 220–33.
Mignon, C. and **Heran, F.** (1979) La Costa del Sol et son arrière-pays, pp. 53–133 in Bernal, A. M. *et al.*, *Tourisme et Développement Régional en Andalousie*, Editions de Boccard, Paris.
Mill, R. C. and **Morrison, A. M.** (1985) *The Tourism System: An Introductory Text*, Prentice-Hall International, Englewood Cliffs.
Milne, S. (1987) *The Economic Impact of Tourism in the Cook Islands*, Occasional Publication 21, Department of Geography, University of Auckland.
Mings, R. C. (1978a) *Climate and Tourism Development: An Annotated Bibliography*, Climatological Publications, Bibliography Series No. 4, State Climatologist for Arizona, Tempe.
Mings, R. C. (1978b) Tourist industry development: at the crossroads, *Tourist Review*, **33**(3), 2–5.
Mings, R. C. (1985) International tourism in Barbados, *Caribbean Geography*, **2**(1), 69–72.
Mings, R. C. (1988) Assessing the contribution of tourism to international understanding, *J. Travel Research*.
Miossec, J. M. (1976) *Eléments pour une Théorie de l'Espace Touristique*, Les Cahiers du Tourisme, C–36, CHET, Aix-en-Provence.
Miossec, J. M. (1977) Un modèle de l'espace touristique, *L'Espace Géographique*, **6**(1), 41–8.

Mitchell, L. S. (1984) Tourism research in the United States: a geographic perspective, *GeoJournal*, **9**(1), 5–15.

Moller, H.-G. (1983) Étude comparée des centres touristiques du Languedoc–Roussillon et de la côte de la Baltique en République Fédérale Allemande, *Norois*, **120**, 545–51.

Molnar, K. and Tozsa, I. (1984) Computerized assessment of touristic potential, pp. 319–26 in Enyedi, G. and Pecsi, M. (eds.), *Geographical Essays in Hungary*, Geographical Research Institute, Hungarian Academy of Sciences, Budapest.

Montgomery G. and Murphy, P. E. (1983) Government involvement in tourism development: a case study of TIDSA implementation in British Columbia, pp. 183–207 in Murphy, P. (ed.), *Tourism in Canada: Selected Issues and Options*, Western Geographical Series, Vol. 21, University of Victoria, Victoria.

Morgan, R. K. (1983) The role of EIA in New Zealand: the present and the future, pp. 160–76 in Heenan, L. D. B. and Kearsley, G. W. (eds.), *Man, Environment and Planning*, Department of Geography, University of Otago, Dunedin.

Mormont, M. (1987) Tourism and rural change: the symbolic impact, pp. 35–44 in Bouquet, M. and Winter, M. (eds.), *Who From Their Labours Rest?*, Avebury, Aldershot.

Morris, A. and Dickinson, G. (1987) Tourist development in Spain: growth versus conservation on the Costa Brava, *Geography*, **72**(1), 16–25.

Mosimann, T. (1985) Geo-ecological impacts of ski piste construction in the Swiss Alps, *Applied Geography*, **5**(1), 29–37.

Murphy, P. E. (1983) Perceptions and attitudes of decision-making groups in tourism centres, *J. Travel Research*, **21**(3), 8–12.

Murphy, P. E. (1985) *Tourism: A Community Approach*, Methuen, New York and London.

Myers, L. W. and Moncrief, P. (1978) Differential leisure travel decision-making between spouses, *Annals of Tourism Research*, **5**(1), 157–65.

National Economic and Social Development Board (1987) *Summary, The Sixth National Economic and Social Development Plan* (1987–91) Bangkok.

National Planning Office (1983) *Republic of Vanuatu: First National Development Plan, 1982–1986*, National Planning Office, Port Vila.

National Planning and Statistics Office (1984) *The Mid-Term Review of Vanuatu's First National Development Plan*, National Planning and Statistics Office, Vanuatu.

Nayacakalou, R. (1972) Investment for tourism in Fijian land, *Pacific Perspective*, **1**(1), 34–7.

Nefedova, V. B., Smirnova, Y. D. and **Schvidchenko, L. G.** (1974) Techniques for the recreational evaluation of an area, *Soviet Geography: Review and Translation,* 15(8), 507–12.

Noronha, R. (1979) Paradise reviewed: tourism in Bali, pp. 177–204 in de Kadt, E. (ed.), *Tourism Passport to Development?,* Oxford University Press, Oxford.

Northland United Council (1986) *Directions for Tourism in Northland,* Northland United Council, Whangarei.

OAS (1978a) *Grenada Tourism Development Plan,* Reports and Studies Series No. 26, Organization of American States, Washington.

OAS (1978b) *Estrategia de Desarrollo Turístico Ecuador,* Serie de Informes y Estudios No. 24, Organization of American States, Washington.

OAS (1978c) *Metodologia de la Investigación Turística en la Planificación del Desarrollo Turístico,* Organization of American States, Washington.

Odouard, A. (1973) Le tourisme et les Iles Canaries, *Les Cahiers d'Outre-Mer, 102,* 150–71.

O'Driscoll, T. J. (1985) European Travel Commission, *Tourism Management,* 6(1), 66–70.

OECD (1967) *Tourism Development and Economic Growth,* Organization for Economic Cooperation and Development, Paris.

OECD (1981a) *The Impact of Tourism on the Environment,* Organization for Economic Cooperation and Development, Paris.

OECD (1981b) *Case Studies of the Impact of Tourism on the Environment,* Organization for Economic Cooperation and Development, Paris.

Oglethorpe, M. K. (1984) Tourism in Malta: a crisis of dependence, *Leisure Studies,* 3(2), 141–61.

O'Grady, R. (1981) *Third World Stopover,* World Council of Churches, Geneva.

Opinion Research Corporation (1980) *A Study of Potential U.S. Vacation Visitors to the Pacific Area: Executive Summary,* PATA, San Francisco.

Oppendijk van Veen, W. M. and **Verhallen, T. W. M.** (1986) Vacation market segmentation: a domain-specific value approach, *Annals of Tourism Research,* 13(1), 37–58.

Papadopoulus, S. I. (1986) The tourism phenomenon: an examination of important theories and concepts, *Tourist Review,* 40(3), 2–11.

Papadopoulus, S. I. and **Mirza, H.** (1985) Foreign tourism in Greece: an economic analysis, *Tourism Management* 6(2), 125–37.

Pavaskar, M. (1982) Employment effects of tourism and the Indian experience, *J. Travel Research,* 21(2), 32–38.

Pawson, I. G. *et al.* (1984) Growth of tourism in Nepal's Everest region: impact on the physical environment and structure of human settlements, *Mountain Research and Development,* 4(3), 237–46.

Pawson, I. G., Stanford, D. D. and **Adams, V. A.** (1984) Effects of modernization on the Khumbu region of Nepal: changes in population structure, 1970–82, *Mountain Research and Development,* 4(1), 73–81.

Pearce, D. G. (1978a) Tourist development: two processes, *Travel Research J.,* 43–51.

Pearce, D. G. (1978b) Skifield development in New Zealand, pp. 91–4 in *Proc. 9th Geog. Conf.,* N.Z. Geog. Soc. Dunedin.

Pearce, D. G. (1978c) A case study of Queenstown, pp. 23–45 in *Tourism and the Environment,* Department of Lands and Survey, Wellington.

Pearce, D. G. (1978d) Form and function in French resorts, *Annals of Tourism Research,* 5(1), 142–56.

Pearce, D. G. (1979a) Towards a geography of tourism, *Annals of Tourism Research,* 6(3), 245–72.

Pearce, D. G. (1979b) Land tenure and tourist development: a review, *Proc 10th Geog. Conf., Auckland.*

Pearce, D. G. (1981) Estimating visitor expenditure, a review and a New Zealand case study, *Int. J. Tourism Management,* 2(4), 240–52.

Pearce, D. G. (1982) Recreation research and policy: implications of the Westland National Park economic impact study, *Tourism Recreation Research,* 7(2) and 8(1), 7–11.

Pearce, D. G. (1983) The development and impact of large-scale tourism projects: Languedoc–Roussillon (France) and Cancun (Mexico) compared, pp. 59–71 in Kissling, C. C. *et al.* (eds.), *Papers, 7th Australian/N.Z. Regional Science Assn,* Canberra.

Pearce, D. G. (1984) Planning for tourism in Belize, *Geographical Review,* 74(3), 291–303.

Pearce, D. G. (1985a) Tourism and planning in the Southern Alps of New Zealand, pp. 293–308 in Singh, T. V. and Kaur, J. (eds.), *Integrated Mountain Development,* Himalayan Books, New Delhi.

Pearce, D. G. (1985b) Tourism and environmental research: a review, *Int. J. Environmental Studies,* 25(4), 247–55.

Pearce, D. G. (1987a) *Tourism Today: a Geographical Analysis,* Longman, Harlow and John Wiley, New York.

Pearce, D. G. (1987b) Spatial patterns of package tourism in Europe, *Annals of Tourism Research,* 14(2), 183–201.

Pearce, D. G. (1987c) Mediterranean charters: a comparative, geographic perspective, *Tourism Management,* 8(4), 291–305.

Pearce, D. G. (1987d) Motel location and choice in Christchurch, *New Zealand Geographer,* 43(1), 10–17.

Pearce, D. G. (1988a) Tourism and regional development in the European Community, *Tourism Management,* 9(1), 13–22.

Pearce, D. G. (1988b) Tourist time-budgets, *Annals of Tourism Research,* 15(1), 109–24.

Pearce, D. G. and **Cant, R. G.** (1981) *The Development and Impact of Tourism in Queenstown,* N.Z. Man and Biosphere Report No. 7, N.Z. National Commission for UNESCO/Department of Geography, University of Canterbury, Christchurch, New Zealand.

Pearce, D. G. and **Grimmeau, J.-P.** (1985) The spatial structure of tourist accommodation and hotel demand in Spain, *Geoforum,* **15**(4), 37–50.

Pearce, D. G. and **Johnston, D. C.** (1986) Travel within Tonga, *J. Travel Research,* **24**(2), 13–17.

Pearce, D. G. and **Kirk, R. M.** (1986) Carrying capacities for coastal tourism, *Industry and Environment,* **9**(1), 3–6.

Pearce, D. G. and **Mings, R. C.** (1984) Geography, tourism and recreation in the Antipodes, *GeoJournal* **9**(1), 91–5.

Pearce, J. A. (1980) Host community acceptance of foreign tourists: strategic considerations, *Annals of Tourism Research,* **7**(2), 224–33.

Pearce, P. L. (1982) *The Social Psychology of Tourist Behavior,* Pergamon, Oxford.

Pearce, P. L. and **Caltabiano, M. L.** (1983) Inferring travel motivation from travellers' experiences, *J. Travel Research,* **22**(2), 16–20.

Peck, J. G. and **Lepie, A. S.** (1977) Tourism and development in three North Carolina coastal towns, pp. 159–172 in Smith, V. (ed.), *Hosts and Guests: The Anthropology of Tourism,* University of Pennsylvania Press, Philadelphia.

Pedevillano, C. and **Wright, R. G.** (1987) The influence of visitors on mountain goat activities in Galcier National Park, Montana, *Biological Conservation,* **39**, 1–11.

Peppelenbosch, P. G. N. and **Tempelman, G.-J.** (1973) Tourism and the developing countries, *Tijdschrift voor Economische en Social Geografie,* **64**(1), 52–8.

Perla, R. and **Glenne, B.** (1981) Skiing, pp. 709–40 in Gray, D. M. and Male, D. H. (eds.), *Handbook of Snow: Principles, Processes, Management and Use,* Pergamon, Toronto.

Perreault, W. D., Darden, D. K. and **Darden, W. R.** (1977) A psychographic classification of vacation life styles, *J. Leisure Research,* **9**(3), 208–24.

Perret, R. and **Bruère, M.** (1970) Les ports du plaisance du Var, *Bulletin du P.C.M.,* mars, 82–90.

Perrin, H. (1971) *Les Stations de Sports d'Hiver,* Berger-Levrault, Paris.

Peters, M. (1969) *International Tourism: The Economics and Development of the International Tourist Trade,* Hutchinson, London.

Phillips, P. H. (1974) Impact reporting: an incremental approach to environmental planning, pp. 63–7 in *Proc. IGU Reg. Conf./8th N.Z. Geog. Conf.,* Palmerston North.

Pigram, J. J. and **Cooper, M. J.** (1980) Economic impact analysis in tourism planning and development, pp. 19–31 in Pearce, D. G.

(ed.), *Tourism in the South Pacific*, N.Z. National Commission for UNESCO/Department of Geography, University of Canterbury, Christchurch.

Piperoglou, J. (1967) Identification and definition of regions in Greek tourist planning, *Papers, Regional Science Association*, 169–76.

Pizam, A. (1978) Tourism's impacts: the social costs to the destination as perceived by its residents, *J. Travel Research*, 16(4), 8–12.

Pizam, A., Neumann, Y. and Reichel, A. (1979) Tourist satisfaction: uses and misuses, *Annals of Tourism Research*, 6(2), 195–7.

Pizam, A, and Pokela, J. (1980) The vacation farm: a new form of tourism destination, pp. 203–16 in Hawkins, D. E., Shafer, E. L. and Rovelstad, J. M. (eds.), *Tourism Marketing and Management Issues*, George Washington University, Washington.

Plog, S. C. (1973) Why destination areas rise and fall in popularity, *Cornell H.R.A. Quarterly*, November, 13–16.

Potter, A. F. (1978) The methodology of impact analysis, *Town and Country Planning*, 46(9), 400–4.

Préau, P. (1968) Essai d'une typologie de stations de sports d'hiver dans les Alpes du Nord, *Revue de Géographie Alpine* 58(1), 127–40.

Préau, P. (1970) Principe d'analyse des sites en montagne, *Urbanisme*, 116, 21–5.

Préau, P. (1983) Le changement social dans une commune touristique de montagne: Saint-Bon-Tarentaise (Savoie), *Revue de Géographie Alpine*, 71(4), 407–29 and 72(2–4), 411–37.

Price, M. F. (1985) Impacts of recreational activities on alpine vegetation in Western North America, *Mountain Research and Development*, 5(3), 263–77.

Priddle, G. and Kreutzwiser, R. (1977) Evaluating cottage environments in Ontario, pp. 165–79 in Coppock, J. T. (ed.), *Second Homes: Curse or Blessing?*, Pergamon, Oxford.

Priestley, G. K. (1986) El turismo y la transformación del territorio: un estudio de Tossa, Lloret de Mar y Blanes a traves de la fotografia aerea 1956–81, pp. 88–106 in *Jornades Técniques Sobre Turisme i Mediambient Sant Feliu de Guixols–Costa Brava*, Barcelona.

Pye, E. A. and Lin, T.-B. (eds.), (1983) *Tourism in Asia: the Economic Impact*, Singapore University Press, Singapore.

van Raaij, W. F. (1986) Consumer research on tourism: mental and behavioural constructs, *Annals of Tourism Research*, 13(1), 1–9.

van Raaij, W. F. and Francken, D. A. (1984) Vacation decisions, activities, and satisfactions, *Annals of Tourism Research*, 11(1), 101–12.

Racey, G. D. and Euler, D. (1983) An index of habitat disturbance for lakeshore cottage development, *J. Environmental Management*, 16, 173–9.

Rajotte, F. (1975) The different travel patterns and spatial framework of recreation and tourism, pp. 43–52 in *Tourism as a Factor in National and Regional Development*, Department of Geography, Trent University, Occasional Paper 4, Peterborough.

Rajotte, F. (1987) Safari and beach-resort tourism: the costs to Kenya, pp. 78–90 in Britton, S. and Clarke, W. C. (eds.), *Ambiguous Alternative: Tourism in Small, Developing Countries*, University of the South Pacific, Suva.

Ranck, S. (1980) The socio-economic impact of recreational tourism on Papua New Guinea, pp. 55–68 in Pearce, D. G. (ed.), *Tourism in the South Pacific: The Contribution of Research to Development and Planning*, N.Z. National Commission for Unesco/Department of Geography, University of Canterbury, Christchurch.

Rao, A. (1986) *Tourism and Export Instability in Fiji*, Occasional Papers in Economic Development, No. 2, Faculty of Economic Studies, University of New England, Armidale.

Reffay, A. (1974) Alpages et stations de sports d'hiver en Haute Tarentaise, *Revue de Géographie Alpine*, **62**(1), 41–73.

Reffay, A. (1985) Alpages et tourisme de part et d'autre du Col du Lautaret, *Revue de Géographie Alpine*, **73**(3), 296–312.

Reime, M. and Hawkins, C. (1979) Tourism development: a model for growth, *Cornell Hotel and Restaurant Administration Quarterly*, **20**(1), 67–74.

Renard, J. (1972) Tourisme balnéaire et structures foncières: l'exemple du littoral vendéen, *Norois*, **73**, 67–79.

Rey, M. (1968) Acondicionamiento del terreno apto para el esqui y equilibro entre la capacidad de la estación y las posibilidades del esqui, pp. 97–112 in *Estaciones para Deportes de Invierno*, Instituto de Estudios Turisticos, Madrid.

Richez, G. (1986) *Tourisme et Attentats Subversifs en Corse*, Paper presented at the meeting of the International Geographical Union's Commission on the Geography of Tourism and Leisure, Palma de Mallorca (mimeo).

Richter, L. K. (1980) The political uses of tourism: a Philippine case study, *J. Developing Areas*, **14**, 237–57.

Richter, L. K. (1985a) Fragmented politics of US tourism, *Tourism Management*, **6**(3), 162–73.

Richter, L. K. (1985b) State-sponsored tourism: a growth field for public administration?, *Public Administration Review*, Nov./Dec. 832–9.

Richter, L. K. and Richter, W. C. (1985) Policy choices in South Asian tourism development, *Annals of Tourism Research*, **12**(2), 201–17.

Ritchie, J. R. B. and Filiatraut, P. (1980) Family vacation decision-making – a replication and extension, *J. Travel Research*, **18**(4), 3–14.

Ritchie, J. R. and Zins, M. (1978) Culture as determinant of the attractiveness of a tourism region, *Annals of Tourism Research,* 5(2), 252–67.

Robertson, R. W. (1977) Second-home decisions: the Australian context, pp. 165–80 in Coppock, J. T. (ed.), *Second Homes: Curse or Blessing?,* Pergamon, Oxford.

Robinson, G. W. S. (1972) The recreation geography of South Asia, *Geographical Review,* 62(4), 561–72.

Robinson, H. (1976) *A Geography of Tourism,* MacDonald and Evans, London.

Rodenburg, E. E. (1980) The effects of scale in economic development: tourism in Bali, *Annals of Tourism Research,* 7(2), 177–96.

Romsa, G. (1981) An overview of tourism and planning in the Federal Republic of Germany, *Annals of Tourism Research,* 8(3), 333–56.

Rose, W. (1981) The measurement and economic impact of tourism on Galveston, Texas: a case study, *J. Travel Research,* 19(4), 3–11.

Rostow, W. W. (1960) *The Stages of Economic Growth,* Cambridge University Press, Cambridge.

Rothman, R. A. (1978) Residents and transients: community reaction to seasonal visitors, *J. Travel Research,* 16(3), 8–13.

Roucloux, J. C. (1977) La demande touristique: ses caractéristiques, sa mesure et son importance, *Travaux Géographiques de Liége,* 165, 63–85.

Ruiz, A. L. (1985) Tourism and the economy of Puerto Rico: an input–output approach, *Tourism Management,* 6(1), 61–5.

Saglio, C. (1979) Tourism for Discovery: a project in Lower Casamance, Sénégal, pp. 321–35 in de Kadt, E. (ed.), *Tourism: Passport to Development?,* Oxford University Press, Oxford.

Saglio, C. (1985) Sénégal: tourisme rural intégré en Basse-Casamance, *Espaces,* 76, 29–32.

Salm, R. V. (1986) Coral reefs and tourist carrying capacity: the Indian Ocean experience, *Industry and Environment,* 9(1), 11–15.

Salvato, J. A. (1976) *Guide to Sanitation in Tourist Establishments,* World Health Organization, Geneva.

Sauran, A. (1978) Economic determinants of tourist demand: a survey, *Tourist Review,* 33(1), 2 4.

Schmidhauser, H. (1975) Travel propensity and travel frequency, pp. 53–60 in Burkart, A. J. and Medlik, S. (eds.), *The Management of Tourism,* Heinemann, London.

Schmidhauser, H. (1976) The Swiss travel market and its role within the main tourist generating countries of Europe, *Tourist Review,* 31(4), 15–18.

Schneider, P., Schneider, J. and **Hansen, E.** (1972) Modernization and development: the role of regional élites and noncorporate groups in the European Mediterranean, *Comparative Studies in Society and History*, **14**(3), 328–50.

Seabrooke, A. K. (1981) The environmental impacts of water-based recreation, *East Lakes Geographer*, **16**, 11–19.

Seers, D. (1969) The meaning of development, *International Development Review*, **11**(4).

Seers, D. (1977) The new meaning of development, *International Development Review*, **19**(3), 2–7.

Seers, D. (1979) The periphery of Europe, pp. 3–34 in Seers, D., Schaffer, B. and Kiljunen, M.-L. (eds.), *Underdeveloped Europe: Studies in Core–Periphery Relations*, Harvester Press, Hassocks.

Seligman, G. (1980) *Snow Structure and Ski Fields*, International Glaciological Society, Cambridge.

Senftleben, W. (1973) Some aspects of the Indian hill stations: a contribution towards a geography of tourist traffic, *Philippines Geographical J.*, **17**(1), 21–9.

Service d'Étude d'Aménagement Touristique du Littoral (n.d.) *Perspectives d'Aménagement à Long Terme du Littoral Français. Equipement et Occupation à Vacation Touristique du Littoral.*, Rapport de synthèse, Paris.

Sessa, A. (1983) *Elements of Tourism Economics*, Catal, Rome.

Seward, A. B. and **Spinard, B. K.** (eds.) (1982) *Tourism in the Caribbean: The Economic Impact*, International Development Research Centre, Ottawa.

Sheldon, P. J. and **Var, T.** (1984) Resident attitudes to tourism in North Wales, *Tourism Management*, **5**(1), 40–7.

Shih, D. (1986) VALS as a tool of tourism market research: the Pennsylvania experience, *J. Travel Research*, **24**(4), 2–11.

Shucksmith, D. M. (1983) Second homes, a framework for policy, *Town Planning Review*, **54**(2), 174–93.

Sibley, R. G. (1982) *Ski Resort Planning and Development*, Foundation for the Technical Advancement of Local Government Engineering in Victoria, Melbourne.

Silverman, G. and **Erman, D. C.** (1979) Alpine lakes in Kings Canyon National Park, California: baseline conditions and possible effects of visitor use, *J. Environmental Management*, **8**(1), 73–87.

Simeral, W. B. (1966) A guide to the appraisal of ski areas, *Valuation*, Sept., 44–61.

Singh, T. V. (1975) *Tourism and Tourist Industry*, New Heights, Delhi.

Singh, T. V. and **Kaur, J.** (1985) In search of holistic tourism for the Himalaya, pp. 365–89 in Singh, R. V. and Kaur, J. (eds.), *Integrated Mountain Development*, Himalayan Books, New Delhi.

Smith, D. M. (1977) *Human Geography: A Welfare Approach*, Arnold, London.

Smith, S. L. J. (1977) Room for rooms: a procedure for the estimation of potential expansion of tourist accommodations, *J. Travel Research*, 15(4), 26–9.

Smith, S. L. J. (1983) *Recreation Geography*, Longman, London.

Smith, V. L. (1977a) Recent search on tourism and culture change, *Annals Tourism Research*, 4(3), 129–34.

Smith, V. L. (ed.) (1977b) *Hosts and Guests: the Anthropology of Tourism*, Pennsylvania Press, Philadelphia.

Smith, V. L., Hetherington, A. and Brumbaugh, M. D. D. (1986) California's Highway 89: a regional tourism model, *Annals of Tourism Research*, 13(3), 415–33.

Smyth, R. (1986) Public policy for tourism in Northern Ireland, *Tourism Management*, 7(2), 120–6.

Social Tourism Study Group (1976) *Holidays: The Social Need*, English Tourist Board, London.

Soesilo, J. A. and Mings, R. C. (1987) The economic impact of winter visitors on Sunbelt cities: revenues and costs of municipal governments, *Visions in Leisure and Business*, 6(3), 40–50.

Spartidis, A. (1976) Employment in the tourist sector and its socio-economic implications, pp. 119–21 in *Planning and Development of the Tourist Industry in the ECE Region*, United Nations, New York.

Stallibrass, C. (1980) Seaside resorts and the holiday accommodation industry: a case study of Scarborough, *Progress in Planning*, 13(3), 103–74.

Stanev, P. (1976) Harmful ecological consequences of the development of the tourist industry and their prevention, pp. 79–82 in ECE, *Planning and Development of the Tourist Industry in the ECE Region*, United Nations, New York.

Stansfield, C. A. (1969) Recreational land use patterns within an American seaside resort, *Tourist Review*, 24(4), 128–36.

Stansfield, C. A. (1973) New Jersey's ski industry: the trend towards market orientation, *J. Travel Research*, 11(3), 6–10.

Stansfield, C. (1978) Atlantic City and the resort cycle: background to the legalization of gambling, *Annals of Tourism Research*, 5(2), 238–51.

Stansfield, C. A. and Rickert, J. E. (1970) The recreational business district, *J. Leisure Research*, 2(4), 213–25.

Stock, R. (1977) Political and social contributions of international tourism to the development of Israel, *Annals of Tourism Research 5* (Special No), 30–42.

Stroud, H. B. (1983) Environmental problems associated with large recreational subdivisions, *Professional Geographer*, 35(3), 303–13.

Sunday, A. A. (1978) Foreign travel and tourism: prices and demand, *Annals of Tourism Research 5*(2), 268–73.

Surrey County Council (1984) *Hotels in Surrey: Report of a Study*, Surrey County Council.

Suzuki, Y. (1967) Tourism in Japan, *Festschrift Leopold G. Scheidl*, Zum 60, Geburtstag, 11, Teil, 204–18.

Symanski, R. and **Burley, N.** (1973) *Tourist Development in the Dominican Republic: an Overview and an Example*, Paper presented at the Conference of Latin American Geographers, Calgary (mimeo).

Takeuchi, K. (1984) Some remarks on the geography of tourism in Japan, *GeoJournal*, **9**(1), 85–90.

Tassin, C. (1984) Tourisme et aménagement régional en Europe de l'Est, *L'Information Géographique*, **48**(5), 188–98 and **49**(1), 26–34.

Taylor, G. D. (1986) Multi-dimensional segmentation of the Canadian pleasure travel market, *Tourism Management*, **7**(3), 146–53.

TDC (Netherlands Institute of Tourism Development Consultants) – SGV – Na Thalang and Co. Ltd (1976) *National Plan on Tourism Development, Final Report*, Tourist Organization of Thailand, Bangkok.

Theuns, D. L. (1976) Notes on the economic impact of international tourism in developing countries, *Tourist Review*, **31**(3), 2–10.

Thirlwall, A. P. (1983) *Growth and Development with Special Reference to Developing Economies*, 3rd edn, Macmillan, London.

Thomason, P., Crompton, J. L. and **Kamp, B. D.** (1979) A study of the attitudes of impacted groups within a host community toward prolonged stay tourist visitors, *J. Travel Research*, **17**(3), 2–6.

Thompson, P. T. (1971) *The Use of Mountain Recreational Resources: a Comparison of Recreation and Tourism in the Colorado Rockies and the Swiss Alps*, University of Colorado, Boulder.

Thomson, C. M. and **Pearce, D. G.** (1980) Market segmentation of New Zealand package tours, *J. Travel Research*, **19**(2), 2–6.

Thurot, J. M. (1973) *Le Tourisme Tropical Balnéaire: le Modèle Caraibe et ses Extensions*, Thesis, Centre d'Études du Tourisme, Aix-en-Provence.

Thurot, J. M. (1980) *Capacité de Charge et Production Touristique*, Études et Memoires No. 43, Centre des Hautes Études Touristiques, Aix-en-Provence.

Thurot, J. M. *et al.* (1976) *Les Effets du Tourisme sur les Valeurs Socio-culturelles*, Centre des Hautes Études Touristiques, Aix-en-Provence.

Tideman, M. C. (1984) Less Dutch holidaymakers in 1982, *Tourist Review*, **38**(2), 28.

Tieh, P. T. H. (1988) *Cultural–Historical Tourism: a Study of Singapore's Chinatown*, Unpublished Ph.D. thesis, Department of Geography, National University of Singapore, Singapore.

Torres Bernier, E. (1985) La construcción de una politica turistica para Andalucia, *Información Commercial Española*, **619**, 109–17.

Tourism Authority of Thailand (1987) *Annual Statistical Report on Tourism in Thailand, 1986,* Tourism Authority of Thailand, Bangkok.

Turner, R. K. (1977) The recreational response to changes in water quality: a survey and critique, *Int. J. Environmental Studies,* **11**(2), 91–8.

United Nations (19. *.eport of the Interregional Seminar on Physical Planning for Tourist Development,* Dubrovnik, Yugoslavia, 19 October–3 November 1970, ST/TAO/Ser. C/131, United Nations, New York.

Unwin, T. (1983) Perspectives on 'development' – an introduction, *Geoforum,* **14**(3), 235–41.

Usher, M. B., Pitt, M. and **de Boer, G.** (1974) Recreational pressures in the summer months on a nature reserve on the Yorkshire Coast, England, *Environmental Conservation,* **1**(1), 43–9.

Uysal, M. and **Crompton, J. L.** (1985) An overview of approaches used to forecast tourism demand, *J. Travel Research,* **23**(4), 7–15.

Var, T., Beck, R. A. D. and **Loftus, P.** (1977) Determination of touristic attractiveness of the touristic areas in British Columbia, *J. Travel Research,* **15**(3), 23-9.

Var, T., Kendall, K. W., Tarakcioglu, E. (1985) Resident attitudes towards tourists in a Turkish resort town, *Annals of Tourism Research,* **12**(4), 652–8.

Var, T. and **Quayson, J.** (1985) The multiplier impact of tourism in the Okanagan, *Annals of Tourism Research,* **12**(4), 497–514.

Vaughan, R. and **Long, J.** (1982) Tourism as a generator of employment: a preliminary appraisal of the position in Great Britain, *J. Travel Research,* **21**(2), 27–31.

Vedenin, Y. A. and **Miroschnichenko, N. N.** (1970) Evaluation of the natural environment for recreational purposes, *Soviet Geography: Review and Translation,* **11**(3), 198–208.

Veyret-Verner, G. (1972) De la grande station à la petite ville: l'exemple de Chamonix-Mont Blanc, *Revue de Géographie Alpine,* **60**(2), 285–305.

Vila Fradera (1966) Trois cas d'actions de l'État pour la localisation touristique dans la nouvelle legislation espagnole, *Tourist Review,* **21**(4), 161–3.

Vogeler, I. (1977) Farm and ranch vacationing, *J. Leisure Research,* **9**(4), 291–300.

Vukonic, B. *et al.* (1978) Italy and Yugoslavia: a case of two touristically advanced countries, pp. 174–204 in *Tourism Planning for the Eighties,* Editions AIEST, Berne.

Wagner, U. (1979) Out of time and place – mass tourism and charter trips, *Ethnos*, **42**(1–2), 38–52.

Wahab, S. E. A. (1973) Tourism and air transport, *Tourist Review*, **28**(4), 146–51 and **29**(1), 9–11.

Wahab, S. E. A. (1975) *Tourism Management*, Tourism International Press, London.

Wahab, S., Crampon, L. J. and **Rothfield, L. M.** (1976) *Tourism Marketing*, Tourism International Press, London.

Wall, G. (1971) Car-owners and holiday activities, pp. 106–7 in Lavery, P. (ed.), *Recreational Geography*, David and Charles, London.

Wall, G., Dudycha, D. and **Hutchinson, J.** (1985) Point pattern analyses of accommodation in Toronto, *Annals of Tourism Research*, **12**(4), 603–18.

Wall, G. *et al.* (1986a) The implications of climatic change for camping in Ontario, *Recreation Research Review*, **13**(1), 50–60.

Wall, G. *et al.* (1986b) Climatic change and recreation resources: the future of Ontario wetlands? *Papers and Proceedings of Applied Geography Conferences*, **9**, 124–31.

Wall, G. and **Sinnott, J.** (1980) Urban recreational and cultural facilities as tourist attractions, *Canadian Geographer*, **24**(1), 50–9.

Wall, G. and **Wright, C.** (1977) *The Environmental Impact of Outdoor Recreation*, Department of Geography Publication Series No. 11, University of Waterloo, Waterloo.

Walter, C. K. and **Tong, H.-M.** (1977) A local study of consumer vacation travel decisions, *J. Travel Research*, **15**(4), 30–4.

Weaver, J. and **Dale, D.** (1978) Trampling effects of hikers, motorcycles and horses in meadows and forests, *J. Applied Ecology*, **15**(2), 451–7.

van der Weg, H. (1982) Revitalization of traditional resorts, *Tourism Management*, **3**(4), 303–7.

Welch, R. V. (1984) The meaning of development: traditional view and more recent ideas, *New Zealand J. Geography*, **76**, 2–4.

Wellings, P. A. and **Crush, J. S.** (1983) Tourism and dependency in Southern Africa: the prospects and planning of tourism in Lesotho, *Applied Geography*, **3**(3), 205–23.

White, J. (1981) *A Review of Tourism in Structure Plans*, Occasional Paper No. 1, Centre for Urban and Regional Studies, University of Birmingham, Birmingham.

White, K. J. and **Walker, M. B.** (1982) Trouble in the travel account, *Annals of Tourism Research*, **9**(1), 37–56.

White, P. E. (1974) *The Social Impact of Tourism on Host Communities: A Study of Language Change in Switzerland*, School of Geography Research Paper 9, University of Oxford, Oxford.

Witter, B. S. (1985) Attitudes about a resort area: a comparison of tourists and local retailers, *J. Travel Research*, **24**(1), 14–19.

Wolfe, R. I. (1970) Discussion of vacation homes, environmental preferences and spatial behaviour, *J. Leisure Research* **2**(1), 85–7.

Wong, P. P. (1986) Tourism development and resorts on the east coast of Peninsular Malaysia, *Singapore J. Tropical Geography*, **7**(2), 152–62.

Wood, R. E. (1979) Tourism and underdevelopment in Southeast Asia, *J. Contemporary Asia*, **9**(3), 274–87.

Wood, R. E. (1984) Ethnic tourism, the state, and cultural change in Southeast Asia, *Annals of Tourism Research*, **11**(3), 353–74.

Woodland, D. J. and **Hooper, J. N. A.** (1977) The effect of human trampling on coral reefs, *Biological Conservation*, **11**, 1–4.

Woodside, A. G. *et al.* (1986) Segmenting the timeshare resort market, *J. Travel Research*, **24**(3), 6–12.

Woodside, A. G. and **Etzel, M. J.** (1980) Impact of physical and mental handicaps on vacation travel behaviour, *J. Travel Research*, **18**(3), 8–11.

Woodside, A. G. and **Jacobs, L. W.** (1985) Step Two in benefit segmentation: learning the benefits realized by major travel markets, *J. Travel Research*, **24**(1), 7–13.

Woodside, A. G. and **Sherrell, D.** (1977) Traveller evoked, inept and inert sets of vacation destinations, *J. Travel Research*, **26**(1), 14–18.

WTO (1979) *Role and Structure of National Tourism Administrations*, World Tourism Organization, Madrid.

WTO (1980a) *Manila Declaration on World Tourism*, World Tourism Organization, Madrid.

WTO (1980b) *Economic Effects of Tourism*, World Tourism Organization, Madrid.

WTO (1980c) *Physical Planning and Area Development for Tourism in the Six WTO Regions, 1980*, World Tourism Organization, Madrid.

WTO (1981) *Tourism Forecasting*, World Tourism Organization, Madrid.

WTO (1983a) *Development of Leisure Time and the Right to Holidays*, World Tourism Organization, Madrid.

WTO (1983b) *Determination of the Importance of Tourism as an Economic Activity Within the Framework of the National Accounting System*, World Tourism Organization, Madrid.

WTO (1983c) *Domestic and International Tourism's Contribution to State Revenue*, World Tourism Organization, Madrid.

WTO (1983d) *Study of Tourism's Contribution to Protecting the Environment*, World Tourism Organization, Madrid.

WTO (1983e) *Concept and Production Innovations of the Tourism Product*, World Tourism Organization, Madrid.

WTO (1984) Tourist carrying capacity, *Industry and Environment*, **7**(1), 30–6.

WTO (1985) *Identification and Evaluation of Existing and New Factors* *and Holiday and Travel Motivations Influencing the Pattern of Present and Potential Domestic and International Tourist Demand,* World Tourism Organization, Madrid.

WTO (1986) *Economic Review of World Tourism,* World Tourism Organization, Madrid.

WTO/Horwath and Horwath (1981) *Tourism Multipliers Explained,* Horwath and Horwath International, London.

WTO–UNEP (1983) *Workshop on Environmental Aspects of Tourism,* World Tourism Organization – United Nations Environmental Programme, Madrid.

Yamamura, J. (1970) Tourism and recreational developments around Tokyo, pp. 63–72 in *Japanese Cities: A Geographical Approach,* Special Publication No. 2, Association of Japanese Geographers, Tokyo.

Yapp, G. A. and **Barrow, G. C.** (1979) Zonation and carrying capacity estimates in Canadian park planning, *Biological Conservation,* 15(3), 191–206.

Yapp, G. A. and **McDonald, N. S.** (1978) A recreation climate model, *J. Environmental Management,* 7, 235–52.

Yokeno, N. (1974) The general equilibrium system of "space-economics" for tourism, *Reports for the Japan Academic Society of Tourism,* 8, 38–44.

Young, R. C. (1977) The structural context of the Caribbean tourist industry: a comparative study, *Economic Development and Cultural Change,* 25(4), 657–72.

Yuill, D., Allen, K. and **Hull, C.** (1980) *Regional Policy in the European Community,* Croom Helm, London.

Zinder, H. (1969) *The Future of Tourism in the Eastern Caribbean,* Zinder and Associates, Washington, D.C.

Author Index

Subject Index

Place Index